Food Engineering Series

Springer's *Food Engineering Series* is essential to the Food Engineering profession, providing exceptional texts in areas that are necessary for the understanding and development of this constantly evolving discipline. The titles are primarily reference-oriented, targeted to a wide audience including food, mechanical, chemical, and electrical engineers, as well as food scientists and technologists working in the food industry, academia, regulatory industry, or in the design of food manufacturing plants or specialized equipment.

More information about this series at http://www.springer.com/series/5996

Ertan Ermiş
Editor

Food Powders Properties and Characterization

Springer

Editor
Ertan Ermiş
Food Engineering Department
Faculty of Engineering and Natural Sciences
Istanbul Sabahattin Zaim University
Istanbul, Turkey

ISSN 1571-0297
Food Engineering Series
ISBN 978-3-030-48907-6 ISBN 978-3-030-48908-3 (eBook)
https://doi.org/10.1007/978-3-030-48908-3

This Springer imprint is published by the registered company Springer Nature Switzerland AG
The registered company address is: Gewerbestrasse 11, 6330 Cham, Switzerland

Preface

Due to recent developments and progress in food powder technology and significant advancement in the analytical and processing possibilities, there has been a gap in the literature in this field. For this reason, we would like to introduce *Food Powders Properties and Characterization* with a great pleasure to our respected readers. The students, industrialists, and researchers studying or dealing with food powders may benefit from this book which presents the fundamental properties of food powders and methods of characterization. The chapters include relevant aspects of particle properties as well as bulk powder properties. The main focus of this book was to give a comprehensive overview of powder characterization and an insight into recent research work related to food powders.

In this book, the physical and chemical properties of food powders and their effect on food powder behaviour are discussed. In addition, some chapters were focused on particle properties, modification of particles, caking–anticaking mechanisms, powder from fruit waste, and microbiological assessment of food powders. We have also included a chapter about rehydration behaviour of food powders which particularly have high protein content. We hope that this book will help to fill the knowledge gap in the literature.

We are very grateful to Springer Nature for their valuable guidance and cooperation. I would like to thank all authors for agreeing to be a part of this book project.

Istanbul, Turkey
April 2020

Ertan Ermiş

Contents

Chapter 1
Food Powders Bulk Properties

Banu Koç, Mehmet Koç, and Ulaş Baysan

1.1 Bulk Density

The bulk density is an important quality criterion during the packaging of powder products and transportation from one place to another. The bulk density also gives information about whether the end product is milled to the desired dimensions or dried to the desired moisture content. Therefore, determining the bulk density of particles means estimation of cost of storage, transportation, product standardization and process success ability in view of the industry.

The most common definition of the bulk density value is a measure of how much powder product can be put into a packaging material having a certain volume. In other words, when a powder just fills a vessel of known volume V, and the mass of the powder is m, then the bulk density of the powder is m/V. After the quantities value of the powder bulk density is determined, it is necessary that these results are evaluated and interpreted.

The particles tend to move towards the bottom of the container depending on time. As time progresses, bulk density increases due to this movement of particles. As a result of an increase in the amount of substance falling to volume per unit (m/V), the density is increased. This change of density is depended on the porosity described the non-occupied volume function (Barbosa-Canovas and Juliano 2005). Thus, bulk density is defined as "the mass of particles that occupies a unit volume of a bed", while porosity is defined as "the volume of the voids within the bed divided by the total volume of the bed". The particle density, which is for a unit

B. Koç (✉)
Gaziantep University, Fine Arts, Gastronomy and Culinary Arts Department, Gaziantep, Turkey
e-mail: banukoc@gantep.edu.tr

M. Koç · U. Baysan
Aydın Adnan Menderes University, Faculty of Engineering, Department of Food Engineering, Aydın, Turkey

E. Ermiş (ed.), *Food Powders Properties and Characterization*, Food Engineering Series, https://doi.org/10.1007/978-3-030-48908-3_1

volume of the powder, is linked to these two properties. Porosity can be a good prediction of the sphericity or irregularity of the particles in a bulk solid. An average porosity calculation of 0.4 or 40% is normal for spheroid particles, whereas irregular shaped or very small particulates have higher porosity values (Woodcock and Mason 1987). High porosity values are a sign of logistical and economic problems that can be encountered during the storage and transportation of powder product (Fasina 2007).

The bulk density of powder is measured as aerated, poured, and tap density considering product type and particle properties (Barbosa-Canovas and Juliano 2005). One of these definitions should be selected, applied and interpreted carefully in view of the process technique and conditions, the usage area of the powder and the structure of the powder. Although each of these definitions has a standard procedure, they are far from universality since interpreting these terms is still confusing. For example, the poured density is the loose bulk density according to some researchers while the apparent density is the poured density in view of others (Fasina 2007). Some researchers evaluated that aerated density is remained the bulk density after the aerated the powder. However, aerated density can be defined as "the particles are separated from each other by a film of air and not being in direct contact with each other". Therefore, each bulk density definition must be well understood before starting the extrapolation.

Poured density is widely used and means to "determine the mass–volume ratio of a powder sample by weighing a container of known volume without the sample and then with the freely poured powder." However, the poured density measurement is modified to any industry or company conditions. This situation causes many difficulties: the same height should be always adjusted for the powder poured; the constant height and diameter vessel should be used etc. Therefore, the measurement of poured bulk density is far from standardized and is specific to each company and conditions.

The powder in the most loosely packed form is defined as the *aerated density*. The particle possessing the dispersed form drops into the measurement cylindrical vessel. Another application is the gas fluidization. The gas fluidization is sometimes used, and gas flow is closed slowly. It is difficult to level the top of the vessel due to the many structure collapsed.

The *tap bulk density* is "the bulk density of a powder that has been settled into closer packing than existed in the poured state by tapping, jolting, or vibrating the measuring vessel." Tapped density is determined by compression of the sample filled in the graduated cylinder. Although tapping can be performed as manual, using the mechanical tapping device is preferable and used since this measurement is standardized approximately and it is possible to repeat the sample preparation conditions.

The definitions described above are intended to determine the porosity. The determination of porosity gives us information on the particle's behavior in bulk. In this case, the porosity and bulk density are one of the most effective parameters on the flow characteristics and behavior of the powder particles. The determination of the flow property is particularly important in determining the size of the packaging material, the volume of the product to be transported in silos, the behavior of the particles in the piping system, and packaging opening required to unpack the product.

1.1.1 The Effect of Process Method and Conditions on Bulk Density

The bulk density of food powders depends on the intensity of attractive inter-particle forces, the air within each particle (occluded air content) and the air between each particle (interstitial air) particle density, particle size, surface activity and degree of adhesion of powder (Barbosa-Canovas and Juliano 2005; Walton 2000). There may be undesirable structural changes in the final powder product such as shrinkage, deformation, expansion, crust formation depending on the evaporation rate during drying. A shell is formed in the droplet during drying and the thickness of the shell varies according to the drying speed. At high drying rates, large grains with thin shells and low density are obtained, while at low drying rates small particles with thick shells and high density are obtained. Depending on the temperature to which the particle is exposed, the water held in the shell during the drying evaporates and forms a pressure towards the shell. As a result, the shell breaks and hollow spheres are obtained. Morphological properties are directly related to bulk properties of powder food products (Schubert 1987). It is known that complex changes in the morphologies (size, shape and appearance) of droplets occur during drying and that the protection of these properties is related to the porosity and integrity of the particles. With respect to morphology, the particles produced by spray drying generally show a smooth surface and are spherical in shape, have lowest surface-to-volume ratio (aroma retention), highest bulk densities (best packing) and best flowability (Kurozawa et al. 2009).

The dry matter content of the material fed to the dryer also affects the morphology of the end product (Koç et al. 2011). Increasing the feed viscosity, by increasing the dry matter content of feed solution or decreasing the feed temperature will cause the formation of larger particles during atomization (Masters 1991; Mujumdar 2007). It has been reported that surface-tension effects during atomization appeared minor, however an increase in feed dry matter content has an effect on the evaporation characteristics where generally there is an increase in bulk density (Masters 1991; Eisen et al. 1998; Mermelstein 2001). The larger particles occupy more pore volume than the smaller particles and provide a decrease in the gap between the particles, hence the higher bulk density of the smaller particles up to a certain diameter (Al-Kahtani and Hassan 1990; Grabowski et al. 2006). The particle shape has also effects on the bulk density of powder products. Because the spherical particles have a low interstitial air content, they have the highest bulk density value at situation which other conditions are kept constant. The bulk density of the powder could be small where the powder is comprised of mainly hollow particles. Thus, the particles, which is a smooth, uniform surface, low size and spherical shape, have greater bulk density so the high bulk density is desirable to reduce transportation and packaging costs (Bicudo et al. 2015). A regular spherical particle shape minimizes the amount of interstitial air. Controlling the amount of occluded air could lead to a higher or lower bulk density. For example, stirring or homogenization of the liquid feed solution may result in the creation of air bubbles inside the liquid, then in the drops and in final powder particles.

Products with low bulk density more prone to oxidation and have low storage stability because of have more space between their particles (Koç et al. 2011). The low bulk density is not desirable due to an increase in the package volume, the cost of packaging material and storage area. Factors, such as moisture content, density, shape and size of the particle, feed rate, powder temperature, feed solid content, atomization type, use of counter-current arrangement for spray dryer directly affect the bulk density (Walton 2000).

Table 1.1 shows schematically some studies evaluating the effect of bulk density from different products and different drying methods/conditions. The moisture content of powder increases, there is a corresponding decrease in bulk density. This is a result of the increase in mass due to moisture gain being lower than the accompanying volumetric expansion of the bulk. Peleg et al. (1973) and Peleg and Moreyra (1979) observed reduced bulk density of water-soluble powders upon increasing moisture content. The reduction in powder bulk density was attributed to the presence of inter-particle liquid bridges which kept them further apart and produced a more-open structure than if the particles were non-cohesive. The moisture content of powder linked to the glass transition temperature. The moisture content reduces the glass transition temperature of powder because of be a plasticizer strongly influencing the glass transition temperature of hydrophilic polymeric products with amorphous regions. The moisture content increase leads to the increasing of molecular mobility and the decreasing of glass transition temperature (Braga et al. 2018). Therefore, the glass transition temperature of powder is other effective factor on bulk density. The bulk densities are also influenced by the surface composition of the powder product, the attraction force between the particles, the surface activity and the degree of adhesion (Fayed and Otten 1997). The bulk density depends on both the particle density and behavior of single particle in the bulk as well as the process methods and conditions.

Bulk density also depends on the carrier agent compositions and ratio and declined with increasing fat content. There are several reports on the decrease in bulk density with increasing the concentration of the carrier agent may be due to increase in feed viscosity and consequent increase in particle size (Bhusari et al. 2014; Goula and Adamopoulos 2004; Fazaeli et al. 2012; Kurozawa et al. 2009; Kha et al. 2010; Yousefi et al. 2010). As seen in Table 1.1, an increase in feed concentration (0–30% maltodestrin or gum arabic) of chicken meat protein hydrolysate powder led to a decrease in powder bulk density. According to Goula and Adamopoulos (2004), increasing the feed concentration generally decreases the bulk density due to the increase in particle size. Tamarind pulp powder with maltodextrin showed the highest bulk density than gum arabic and whey protein concentrate. The bulk

Table 1.1 Effects of different drying methods/conditions on bulk density

Food powder	Drying method	Drying conditions	Moisture content (%)	Bulk density (kg/m^3)	Tapped bulk density (kg/m^3)	Main Results	References
Chicken meat protein hydrolysate powder	Spray	Carrier agent: Maltodextrin (10 DE) and gum arabic concentration: 10%, 20% and 30% (w/w) Two fluid nozzle T_{inlet}: 180 °C T_{outlet}: 90–102 °C V_{feed}: 300–500 mL/min	Without carrier agent-1.8 ± 0.1 MD10-1.5 ± 0.1 MD20-1.4 ± 0.1 MD30-1.2 ± 0.1 GA10-1.7 ± 0.1 GA20-1.5 ± 0.1 GA30-1.2 ± 0.1	Without carrier agent-383 ± 7.2 MD10-330 ± 6.1 MD20-305 ± 1.6 MD30-295 ± 2.5 GA10-330 ± 13.3 GA20-311 ± 7.5 GA30-295 ± 8.2	–	The bulk density reduced with increasing in feed concentration (0–30% MD or GA)	Kurozawa et al. (2009)
Gac fruit aril powder	Spray	Carrier agent: Maltodextrin (12 DE) concentration: 10%, 20% and 30% (w/w) Two fluid nozzle T_{inlet}: 120, 140, 160, 180 and 200 °C T_{outlet}: 83, 94, 103, 112 and 120 °C V_{feed}: 12–14 mL/min	MD10-4.87 ± 0.71 MD20-4.54 ± 0.54 MD30-4.06 ± 0.47 DT-120-5.29 ± 0.50 DT-140-4.81 ± 0.49 DT-160-4.47 ± 0.48 DT-180-4.01 ± 0.18 DT-200-3.88 ± 0.35	720 ± 0.05 700 ± 0.06 730 ± 0.07 780 ± 0.05 740 ± 0.05 700 ± 0.03 690 ± 0.05 660 ± 0.04	–	The bulk density of Gac powders was significantly affected by the drying temperature, with decreasing density observed with increased drying temperature	Kha et al. (2010)

(continued)

Table 1.1 (continued)

Food powder	Drying method	Drying conditions	Moisture content (%)	Bulk density (kg/m^3)	Tapped bulk density (kg/m^3)	Main Results	References
Microencapsulated rosemary oil powder	Spray	Wall material gum arabic concentration: 10–30% (w/v) Two fluid nozzle T_{inlet}: 135–195 °C V_{feed}: 0.5–1 L/h CCRD design	In the range of 0.26–3.16%	In the range of 250–360	In the range of 410–520	The bulk density was influenced positively by the wall material concentration and negatively by the inlet air temperature, flow rate, and air temperature interaction whereas tapped density was significantly influenced by the temperature variable only	de Barros Fernandes et al. (2013)
Tamarind powder	Spray	Carrier agent: Maltodextrin (20 DE), gum arabic and whey protein concentrate concentration: 40, 50 and 60% for maltodextrin MD and GA and 10, 20 and 30% for WPC T_{inlet}: 180 °C T_{outlet}: 80 °C V_{feed}: 600 mL/h	MD40-7.11 MD50-6.00 MD60-4.48 GA40-5.60 GA50-4.54 GA60-3.65 WPC10-5.04 WPC20-6,58 WPC30-7.15	MD40-685 MD50-594 MD60-503 GA40-658 GA50-568 GA60-490 WPC10-492 WPC20-467 WPC30-391	–	Bulk density of tamarind powder decreased with increase in addition rate of the carrier agent	Bhusari et al. (2014)

Microencapsulation Sunflower oil and palm oil	Spray	Two fluid nozzle Counter current spray drying water evaporation rate: 20 L/h T_{inlet}: 185 °C T_{outlet}: 80 or 90 °C	SO-L: 2.27 SO-H: 1.77 PO-L: 2.29 PO-H: 1.81	SO-L: 410 SO-H: 350 PO-L: 400 PO-H: 340	–	The powders produced at a higher outlet temperature had a lower bulk density than powders produced at the lower outlet temperature for all powders, regardless of oil type used	Kelly et al. (2014)
Maltodextrin	Spray	T_{inlet}: 140 and 200 °C T_{feed}: 10 and 50 °C V_{feed}: 2.1E−04 and 9.6E−04 kg s^{-1} $V_{air\text{-}flow}$: 1.3E−0.4 and 1.9E−0.4 m^3 s^{-1}	3.813–11.13	245.2–349.1	402.4–572.9	Bulk density was affected by all the independent variables except atomizing air flow whereas tapped density was affected by only inlet air temperature and feed flow rate, respectively	Koç and Kaymak-Ertekİn (2014)

(continued)

Table 1.1 (continued)

Food powder	Drying method	Drying conditions	Moisture content (%)	Bulk density (kg/m^3)	Tapped bulk density (kg/m^3)	Main Results	References
Pink Guava powder	Spray	Carrier agent: Maltodextrin (10 DE) concentration:10%, 15% and 20% (w/v) Two fluid nozzle T_{inlet}: 150, 160 and 170 °C V_{feed}: 350 mL/h	MD10%: 3.34, 3.07 and 2.59 MD15%: 3.18, 3.02, and 2.48 MD20%: 2.96, 2.75 and 2.32	MD10%: 403, 377 and 342 MD15%: 449, 433 and 423 MD20%: 428, 418 and 395	MD10%: 479, 458 and 421 MD15%: 516, 503 and 491 MD20%: 503, 495 and 483	The bulk and tapped densities reduced with increasing inlet temperature The bulk and tapped densities increased with the increasing of MD concentrations except for 15% MD concentration that showed maximum values	Shishir et al. (2014)
Jamun fruit juice powder	Spray	Carrier agent: Maltodextrin (20 DE, 1:4 w\v) Two fluid nozzle T_{inlet}: 140–170 °C T_{outlet}: 80 °C V_{feed}: 10 mL/min	In the range of 3.22 ± 0.09–4.18 ± 0.09	In the range of 240 ± 0.02–260 ± 0.03	In the range of 380 ± 0.02–480 ± 0.03	Jamun juice powder produced at different inlet temperatures showed nonsignificant difference in bulk density whereas tapped density values showed significant difference	Santhalakshmy et al. (2015)

Microencapsulated extra virgin olive oil powder (MEVOP)	Spray	Carrier agent: Maltodextrin and WPC concentration: 0–100% $V_{homogenization}$: 10,000–20,000 rpm Two fluid nozzle T_{inlet}: 200 °C V_{feed}: 5–8 mL/min D-optimal mixture design	In the range of 0.41–2.54%	In the range of 205–530	In the range of 403–761	The bulk density of MEVOP was negatively influenced by the moisture. The bulk and tapped densities were significantly affected by the wall material composition	Koç et al. (2015)
Watermelons powder	Spray	Carrier agent: Maltodextrin Two fluid nozzle T_{inlet}: 120, 130,140 and 150 °C T_{outlet}: 85 °C	2.09 ± 0.023 1.98 ± 0.45 1.78 ± 0.11 1.43 ± 0.044	460 ± 0.01 450 ± 0.01470 ± 0.02430 ± 0.01	–	The bulk densities of watermelon powder unaffected by the inlet temperature	Yue et al. (2018)

(continued)

Table 1.1 (continued)

Food powder	Drying method	Drying conditions	Moisture content (%)	Bulk density (kg/m^3)	Tapped bulk density (kg/m^3)	Main Results	References
Seed gum	Oven Vacuum oven Spray Freeze	Homogenized-gum solution: 10% (w/v) dried sample milling: 1.0 mm sieve 105 °C for 3 h 60 °C for 24 h at 5 psi Centrifugal atomizer T_{inlet}: 160 °C T_{outlet}: 80–85 °C $P_{atomization}$: 552 kPa V_{feed}: 50 mL/min −20 °C for 24 h then −40 °C for 48 h	–	203 195 179 173	253 244 206 197	The drying process significantly influenced the bulk and tapped densities. The freeze-dried gum showed the lowest bulk and tapped densities whereas the oven-dried sample had the highest values	Mirhosseini and Amid (2013)

Coffee	Spray-freeze Spray Freeze	Nozzle: Two fluid nozzle V_{feed}: 6 mL/min Main drying: −25 to −10 °C 107 Pa Secondary drying: 10 °C 40 Pa Two fluid nozzle T_{inlet}: 150 °C T_{outlet}: 100 °C T_{shelf}: 40–10 °C	8.665 ± 0.001 5.347 ± 0.498 8.847 ± 0.129	612 ± 0.007 328 ± 0.002 345 ± 0.006	679 ± 0.008 388 ± 0.001 361 ± 0.004	The FD coffee had a lower bulk density than the SFD sample. The reduced drying temperature during SFD might have resulted in a simultaneous increase in bulk density and solubility of the coffee powder	Ishwarya et al. (2015)

(continued)

Table 1.1 (continued)

Food powder	Drying method	Drying conditions	Moisture content (%)	Bulk density (kg/m^3)	Tapped bulk density (kg/m^3)	Main Results	References
Microencapsulated vitamin E powder	Spray-freeze Spray Freeze	Carrier agent: WPC (1:3 w/v) Nozzle: Two fluid nozzle Main drying: −25 to −10 °C at 0.8 Torr Secondary drying: 10 °C 0.3 Torr Two fluid nozzle T_{inlet}: 100 °C T_{outlet}: 80 °C V_{feed}: 4 mL/min − 25 °C for 2 h then T_{shelf}: −25 to 20 °C Main drying: 8–18 h at 7.6E−2 to 0.8 Torr Secondary drying: 2 h at 25 °C	5.41 ± 0.24 6.99 ± 0.21 7.16 ± 0.52	266 ± 2.40 266 ± 2.40 227 ± 1.69	321.34 ± 2.57 513.26 ± 6.69 280.27 ± 2.79	Among the formulations, FD and SFD microcapsules exhibited comparatively lower bulk density due to more external voids and resulted in higher bulk volume	Parthasarathi and Anandharamakrishnan (2016)

Microbial transglutaminase	Spray-freeze Freeze	Nozzle: ultrasonic, 48 kHz V_{feed}: 6.37 mL/min Main drying: 6 h at 1 mbar Secondary drying: 2 h at 0.01 mbar −80 °C for 4 h then Main drying: 6 h at 1 mbar Secondary drying: 2 h at 0.01 mbar	7.04 (±0.08) 8.64 (±0.33)	152.30 (±0.08) 118.26 (±2.37)	244.32 (±1.50) 231.76 (±6.07)	Freeze dried powder had relatively low bulk and tapped densities which was associated with high porosity and the higher moisture content. The larger particle size and irregular shape of the freeze-dried particles and a consequent increase in inter particle voids with smaller contact surface areas per unit volume may have led to a lower bulk density than those of the spray-freeze dried sample	Isleroglu et al. (2018)

(continued)

Table 1.1 (continued)

Food powder	Drying method	Drying conditions	Moisture content (%)	Bulk density (kg/m^3)	Tapped bulk density (kg/m^3)	Main Results	References
Maltodextrin	Spray-freeze	Nozzle: ultrasonic, 48 kHz V_{feed}: 8 mL/min T_{shelf}: 25–45 °C Main drying: 6–16 h at 1 mbar Secondary drying: 2 h at 0.01 mbar CCRD design	In the range of 2.20 (±0.05)–3.18 (±0.13)	In the range of 71.3 (±0.1)–89.6 (±3.9)	In the range of 118.9 (±1.2)–138.0 (±4.5)	Bulk and tapped densities were affected by all the independent variables	Türker et al. (2018)

SO sunflower oil, *PO* palm oil, Suffix L and H correspond to lower (80 °C) and higher (90 °C) outlet temperatures for spray drying, *CCRD* central composite rotatable design, *FD* freeze drying, *SFD* spray freeze drying, *SD* spray drying, *MD* maltodextrin, *GA* gum arabic, *WPC* whey protein concentrate, *DT* drying temperature, *DE* dextrose equivalent density of MEVOP increased with increasing whey protein concentrate ratio in wall material combinations due to higher particle size and the highest bulk density had the MEVOP with highest maltodextrin ratio. The heavier material accommodates itself more easily in the spaces among the particles, resulting in higher bulk density (Tonon et al. 2010). Similarly, the higher porosity or lower bulk density in spray-dried mango powder was observed due to the addition of maltodextrin and an increase in maltodextrin dextrose equivalent leads to an increase in bulk density. This can be attributed to the fact that the higher the maltodextrin DE, the lower its glass transition temperature (Adhikari et al. 2004; Goula and Adamopoulos 2010; Fazaeli et al. 2012). Shrestha et al. (2007) demonstrated that an increase in maltodextrin concentration causes a decrease in bulk density of orange juice powder. Goula and Adamopoulos (2010) also explained that maltodextrin is considered a skin-forming material and by using it as carrier can induce accumulation and trapping of air inside the particle causing it to become less dense and porous. On the contrary, some studies reported an increase in the bulk density of the final product with an increase in the carrier agent (Sablani et al. 2008; Miravet et al. 2015; Nadeem et al. 2011). The type and properties of the carrier material have a significant effect on the bulk density of powders.

The atomizer is the most important feature of a spray dryer. Its selection and operation are of great importance in achieving cost-efficient production while maintaining product quality (Masters 1991). It is expected to obtain lower bulk densities at higher atomization pressure since producing of smaller particles by effective atomization. Decrease in the bulk density with the increase of the atomization speed was a result of the particle size and the moisture content of the samples. Rotary atomization generally produces a larger particle size in comparison to nozzle atomization. Two-fluid nozzle atomizers obtain the smallest particles sizes.

There are controversial reports on the effect of drying temperature on the bulk density of the powder obtained by spray dryer. Generally, an increase in the inlet temperature often causes a reduction in bulk density, as evaporation rates are faster and products dry to a more porous or fragmented structure (Eisen et al. 1998; Mujumdar 2007). The effect of inlet temperature on bulk density is depicted in Table 1.1. The bulk density of Gac powders, black mulberry powders and pink guava powders was significantly affected by the drying temperature with decreasing density observed with increased drying temperature. This is consistent with the findings of a number of studies, that increasing inlet air drying temperature results in reducing bulk density (Walton and Mumford 1999; Cai and Corke 2000; Goula et al. 2004; Chegini and Ghobadian 2005). At very high temperatures, very high drying processes are achieved implying a lower shrinkage of the droplets, and so a lower density of the powder (Walton 2000; Chegini and Ghobadian 2005). Jamun juice powder produced at different inlet temperatures showed nonsignificant difference in bulk density. The highest bulk density was shown at an inlet temperature of 155 °C, whereas the lowest bulk density was shown at an inlet temperature of 150 °C. Kelly et al. (2014) reported that the powders produced at a higher outlet temperature had a lower bulk density than powders produced at the lower outlet temperature for all powders, regardless of oil type used.

The same material could, depending on the drying methods, result in two powders with completely different bulk densities. Freeze dried powders have low bulk density resulting from the needlelike void spaces that were previously occupied by the ice crystals. Isleroglu et al. (2018) reported that freeze dried powder had relatively low bulk and tapped densities which was associated with high porosity. The larger particle size and irregular shape of the freeze-dried particles and a consequent increase in inter particle voids with smaller contact surface areas per unit volume may have led to a lower bulk density than those of the spray-freeze dried sample. Similar finding was reported by Ishwarya and Anandharamakrishnan (2015), the larger particle size of freeze-dried coffee than the spray-freeze dried coffee. In this study they demonstrated that the higher tapped bulk densities of spray-freeze dried coffee when compared to that of spray dried sample. Caparino et al. (2012) investigated the influence of four drying methods (Refractance Window® drying (RW), freeze drying, drum drying and spray drying) on bulk density of mango powder. They were reported that freeze and spray dried mango powders had significantly lower bulk densities and higher porosities compared to drum and RW dried products.

1.2 Flowability

The flow of powder results from movement of a single particle and the bulk movement of particles. The particle flow takes place on the surface of the other particles in the bulk or on the wall surface of the container (Peleg 1977). Determination of the flow characteristics of the powder, which is quantitative and qualitative identification, provides information on the design of equipment and performance estimation (Sutton 1976). The flow characteristics of the powder products are the great importance for the transportation and storage process in bulk (Chen 1994). Storage and transportation from one place to in other place for product having fluid property does not lead to a serious difficulty. The handling fluids from one point to another is quite easy, however the handling of powder is more difficult due to cohesion between the particles, friction on particle surface and adhesion on the container wall surface. The high flowability of powder particles have spherical shape, smooth surface, large diameters and no agglomeration whereas, the low flowability of powder particles have high stickiness, hackly surface and away from sphericity (Walton 2000). The main forces affected on the flowability of powder are gravity, friction, cohesion and adhesion. The cohesion force is attraction between the particles while, the adhesion force is attraction between particle and wall surface. Moreover, particle composition and properties (size, shape, density and morphology) is effective factors on flowability of powder. In general, materials with narrow particle size distribution have a better flow than materials with wider particle size distribution (Benković et al. 2012). Also, it is generally considered that materials with particle sizes larger than 200 μm are free flowing, while fine powders with particle sizes less than 200 μm are subject to cohesion and flowability problem

(Fitzpatrick 2005; Teunou et al. 1999; Fitzpatrick 2007). The particle surface area increases with the reduction of particle size. This being the case, an increase in the sticky structure may also occur with respect to the moisture content, so flowability problems arise in the powder. As mentioned above, the sticky structure is the result of the cohesion forces on surface of the particles as a result of the bridges they have built and the adhesion forces between particles and the container wall. An increase in the moisture content of powder causes an increase in the degree of cohesion and adhesion force. Thus, the powder flow is affected adversely (Johanson 2005). Furthermore, the glass transition temperature of powder is considered as an important property to evaluate the degree of stability of the product during long term storage periods. In appropriate drying, sticky product can occur if the drying temperature is increased to above the glass transition temperature of the feed material. Especially, sugar and acid-rich powder, which have the low glass transition temperature, tend to stick to the surfaces on which they come into contact such as dryer, container wall etc. and stick to each other during the process and storage period. Thus, a structure like cake is obtained instead of free-flowing powder and the powder flow get worse (Roos 2003).

It is necessary that food industry get information in terms of the size of the packaging material, the volume of the product to be transported in silos, the behavior of the particles in the piping system, and packaging opening required to unpack the product. Therefore, food industry can use some characteristics of angle in order to evaluate aspects of storage and transport. One of the most preferred and used definition is the angle of repose since it is practical, cheap and standardized technique. The static **angle of repose** is defined as "*the angle at which a material will rest on a stationary heap; it is the angle formed by the heap slope and the horizontal when the powder is dropped on a platform.*" Angle of repose values below 35^0 were classified as free flowing; values between 35 and 45^0 indicated some cohesiveness, while values greater than 55^0, showed high cohesiveness with tendency to cause flow problems (Peleg 1977; Chang et al. 1998). The effect of particle size, shape and moisture content is the important on the angle of repose value. A decrease in the angle of repose results from an increase in particle size since smaller particles tend to adhere each other (Teunou et al. 1995).

The flowability is generally expressed by the Carr index as determined by (Carr 1965) as a function of bulk and tapped density in the literature. If Carr index value is less than 15, the flowability of particle is considered to be very good, between 15 and 20 good, between 20 and 35 weak, between 35 and 45 bad, and above from 45 very bad. Furthermore, the powder flow can also be explained by the concept of compressibility. For this, the expression of Hausner rate is used. Hausner (1967) stated that the Hausner ratio (HR) can be calculated by proportioning the tapped density of the powder to the bulk density. The powder product has low stickiness when the HR value is less than 1.2; 1.2–1.4 medium; it is defined as the high stickiness when greater than 1.4. As a result, Carr index and Hausner ratio definitions have allowed to standardize the flowability values of powder products.

1.2.1 The Effect of Process Method and Conditions on Flowability

Several factors affect powder flowability, like size, shape and composition of the surface of the particles (Teunou et al. 1999; Fitzpatrick et al. 2004). Teunou et al. (1995) found that the angle of repose decreases with increasing particle size mainly because smaller particles tend to adhere much more strongly to each other. Particle size has a major influence on powder flowability. Large mean particle size, narrow particle size distribution, spherical shape and smooth surfaces with no sticky or fat components contribute to a better flowability. Liu et al. (2008) reported that an increase in particle size causes a reduction in the Hausner ratio. This is an indication of improved flowability with increase in particle size. Moisture content is an important variable that also affects cohesive strength of bulk solids (Johanson 1978) during storage with cohesion generally increasing with moisture content Fitzpatrick et al. (2004). Teunou et al. (1999) stated that the strength of liquid bridges formed between particles depends on moisture adsorption. Chang et al. (1998) investigated that an increase in Hausner Ratio, angle of repose, and shear stress (all of which indicate a decrease in flowability) as the moisture content of their food powders increased. Zou and Yu (1996) observed a decrease in Hausner ratio with an increase in sphericity. On the other hand, irregular shaped powder exhibited lower flowability, higher Hausner ratios, higher interlocking between the particles and higher coefficient of internal friction. This behavior of irregular particles explained as due to interlocking between particles, which prevents their motion and hence increases interparticle friction during powder flow (Chan and Page 1997).

Table 1.2 shows schematically some studies evaluating the effect of flowability from different products and different drying methods/conditions. Gallo et al. (2011) found that the angle of repose was affected by atomization air flow rate and solids concentration. This value was improved when the atomization air flow rate was decreased and the solids concentration was increased. Both actions led to bigger particles due to greater droplet sizes with higher solids content. Yue et al. (2018) showed that the repose angle of the powder spray dried at inlet temperature of 120 °C was significantly lower than that dried at all other temperatures and the authors also reported that the repose angle of the powder dried at 150 °C was significantly higher than that of the powder dried at 120 °C although their particle sizes were similar and they explained this difference due to the higher moisture content of the powder dried at 120 °C. The significant effect of the drying process on the angle of repose was reported by Nep and Conway (2011). Similar results were also found by Mirhosseini and Amid (2013). The authors stated that the oven-dried gum exhibited the highest angle of repose; while the spray dried gum and freeze-dried gum showed the lowest angle of repose among all dried samples (Table 1.2). This situation could be explained by the oven drying at the elevated temperature (105 °C) might have caused the collapse in the gum structure and the thermal degradation possibly induced by high drying temperature might result in more compact and rigid powder with the low porosity (Mirhosseini and Amid 2013). Ishwarya and

Table 1.2 Effects of different drying methods/conditions on flowability

Food powder	Drying method	Drying conditions	Moisture content (%)	Particle size (μm)	angle of repose (°)	Carr index (%)	Hausner ratio	Main results	References
Rhamnus purshiana extract powder	Spray	Carrier agent: colloidal silicon dioxide ratio: 0.5:1 and 1:1 Two fluid nozzle T_{inlet}: 130–170 °C T_{outlet}: 44–96 °C $V_{air\text{-}flow}$: 400–800 L/h V_{feed}: 1–3 mL/min Feed concentration: 5.59–7.32% (w/w) 2^{5-1} factorial design	In the range of 2.41 ± 0.08–4.72 ± 0.28	In the range of 7.94 ± 0.08–14.43 ± 0.85	In the range of 27–36	In the range of 16.84 ± 2.70–28.23 ± 2.78	–	High solids concentration, high carrier content and low atomization air flow rate led to powders with good flowability properties	Gallo et al. (2011)
Rosemary oil powder		Wall material gum arabic concentration: 10–30% (w/v) Two fluid nozzle T_{inlet}: 135–195 °C V_{feed}: 0.5–1 L/h CCRD design	In the range of 0.26–3.16%	–	–	In the range of 23.09–40.22%	In the range of 1.30–1.67	Both HR and CI values influenced significantly only by the temperature	de Barros Fernandes et al. (2013)

(continued)

Table 1.2 (continued)

Food powder	Drying method	Drying conditions	Moisture content (%)	Particle size (μm)	angle of repose (°)	Carr index (%)	Hausner ratio	Main results	References
Pink Guava powder	Spray	Carrier agent: Maltodextrin (with 10 DE) concentration: 10%, 15% and 20% (w/v) Two fluid nozzle T_{inlet}: 150, 160 and 170 °C V_{feed}: 350 mL/h	MD10%: 3.34, 3.07 and 2.59 MD15%: 3.18, 3.02, and 2.48 MD20%: 2.96, 2.75 and 2.32	MD10%: 12, 12.5 and 13.2 MD15%: 10.4, 11.0, and 12.9 MD20%: 12.9, 13.6 and 14.0	–	MD10%: 1.20, 1.23 and 1.24 MD15%: 1.15, 1.16 and 1.18 MD20%: 1.20, 1.21 and 1.22	MD10%: 17.3, 19.0 and 20.3 MD15%: 13.4, 14.3 and 115.6 MD20%: 17.1, 17.9 and 18.6	HR and CI values increased with the increasing of temperature and MD concentration The 15% MD samples showed lower HR and CI values that indicated good flowability. There is a possibility that for lower particle size and higher bulk density that led to good flowability	Shishir et al. (2014)
Tamarind powder	Spray	Carrier agent: Maltodextrin (with 20 DE, gum arabic and whey protein concentrate concentration: 40, 50 and 60% for maltodextrin MD and GA and 10, 20 and 30% for WPC T_{inlet}: 180 °C T_{outlet}: 80 °C V_{feed}: 600 mL/h	MD40-7.11 MD50-6.00 MD60-4.48 GA40-5.60 GA50-4.54 GA60-3.65 WPC10-5.04 WPC20-6,58 WPC30-7.15	–	–	MD40-1.42 MD50-1.28 MD60-1.34 GA40-1.52 GA50-1.37 GA60-1.29 WPC10-1.29 WPC20-1.23 WPC30-1.28	MD40-29.83 MD50-21.83 MD60-24.41 GA40-34.16 GA50-28.74 GA60-24.27 WPC10-21.97 WPC20-19.34 WPC30-20.47	Tamarind powder with 20% WPC which showed medium flowing character, which may be due to its large particle size and intermediate moisture content	Bhusari et al. (2014)

Food powder	Drying method	Drying conditions	Moisture content (%)	Particle size (μm)	angle of repose (°)	Carr index (%)	Hausner ratio	Main results	References
Jamun fruit juice powder	Spray	Carrier agent: Maltodextrin (with 20 DE (1:4 w\v) Two fluid nozzle T_{inlet}: 140–170 °C T_{outlet}: 80 °C V_{feed}: 10 mL/min	In the range of 3.22 ± 0.09–4.18 ± 0.09	–	–	In the range of 36.10 ± 2.98–41.58 ± 4.51	In the range of 1.57 ± 0.08–1.72 ± 0.14	The spray-dried jamun juice powder had similar flow characteristics and were considered as very cohesive powder. The highest values HR and CI were shown by sample dried at 170 °C, while the lowest were shown by sample dried at 140 °C Hausner ratio and Carr index values were affected by inlet air temperature	Santhalakshmy et al. (2015)
Watermelons powder	Spray	Carrier agent: Maltodextrin Two fluid nozzle T_{inlet}: 120, 130, 140 and 150 °C T_{outlet}: 85 °C	2.09 ± 0.023 1.98 ± 0.45 1.78 ± 0.11 1.43 ± 0.044	21.64 ± 1.22 18.21 ± 0.22 13.44 ± 0.36 21.21 ± 0.26	33.8 ± 0.52 41.5 ± 0.84 43.3 ± 0.96 45.4 ± 0.90	–	–	The repose angle of the powder dried at 120 °C was significantly lower than that dried at all other temperatures because of the higher moisture content	Yue et al. (2018)

(continued)

Table 1.2 (continued)

Food powder	Drying method	Drying conditions	Moisture content (%)	Particle size (μm)	angle of repose (°)	Carr index (%)	Hausner ratio	Main results	References
Seed gum	Oven Vacuum oven Spray Freeze	homogenized-gum solution:10% (w/v) dried sample milling: 1.0 mm sieve 105 °C for 3 h 60 °C for 24 h at 5 psi Centrifugal atomizer T_{inlet}: 160 °C T_{outlet}: 80–85 °C $P_{atomization}$:552 kPa V_{feed}: 50 mL/min −20 °C for 24 h then −40 °C for 48 h	–	–	42.22 35.00 31.50 30.83	–	–	The oven-dried gum exhibited the highest angle of repose; while the spray-dried gum and freeze-dried gum showed the lowest angle of repose among all dried samples	Mirhosseini and Amid (2013)
Coffee	Spray-freeze Spray Freeze	Nozzle: Two fluid nozzle V_{feed}: 6 mL/min Main drying: −25 to −10 °C at 107 Pa Secondary drying: 10 °C at 40 Pa Two fluid nozzle T_{inlet}: 150 °C T_{outlet}: 100 °C T_{shelf}: 40–10 °C	8.665 ± 0.001 5.347 ± 0.498 8.847 ± 0.129	91.1 50.41 636.8	–	10 ± 0.0001 15.5 ± 0.707 4.5 ± 0.707	1.11 ± 0.0001 1.18 ± 0.009 1.05 ± 0.008	The flow of SFD and SD were in the medium range and that of FD in the free flowing zone the flow improved with an increase in particle size in the order of SD < SFD < FD	Ishwarya et al. (2015)

Food powder	Drying method	Drying conditions	Moisture content (%)	Particle size (μm)	angle of repose (°)	Carr index (%)	Hausner ratio	Main results	References
Micro-encapsulated vitamin E powder	Spray-freeze Spray Freeze	Carrier agent: WPC (1:3 w/v) Nozzle: Two fluid nozzle Main drying: −25 to −10 °C at 0.8 Torr Secondary drying: 10 °C 0.3 Torr Two fluid nozzle T_{inlet}: 100 °C T_{outlet}: 80 °C V_{feed}: 4 mL/min − 25 °C for 2 h then T_{shelf}: −25 to 20 °C Main drying: 8–18 h at 7.6E−2 to 0.8 Torr Secondary drying: 2 h at 25 °C	5.41 ± 0.24 6.99 ± 0.21 7.16 ± 0.52	145.3 ± 65.5 195.8 ± 46.6 279.0 ± 23.2	–	16 32 20	1.19 1.47 1.25	SFD microcapsules had 'fair' flowability than SD and FD microcapsules	Parthasarathi and Anandharamakrishnan (2016)
Maltodextrin	Spray-freeze	Nozzle: ultrasonic, 48 kHz V_{feed}: 8 mL/min T_{shelf}: 25–45 °C Main drying: 6–16 h at 1 mbar Secondary drying: 2 h at 0.01 mbar CCRD design	In the range of 2.20 (±0.05)–3.18 (±0.13)	–	–	In the range of 30.0 (±0.3)–40.4 (±0.6)	In the range of 1.43 (±0.01)–1.68 (±0.02)	Hausner ratio and Carr index values were affected by all the independent variables	Türker et al. (2018)

CCRD central composite rotatable design, *HR* Hausner ratio, *CI* Carr index, *MD* maltodextrin, *GA* gum arabic, *WPC* whey protein concentrate, *FD* freeze drying, *SFD* spray freeze drying, *SD* spray drying

Anandharamakrishnan (2015) observed the Hausner ratio value decreased with an increase in particle size in the order of SD < SFD < FD. Smaller mean particle size as in spray freeze dried and spray dried samples indicated the presence of more fines in the powder bulk. The higher free flow values of spray freeze dried coffee when compared to that of spray dried can be due to its higher residual moisture content, non-cohesive nature (Geldart 1973) and wider particle size distribution than the spray-dried powder. The larger particle size of freeze-dried coffee and a consequent increase in interparticle voids with smaller contact surface areas per unit volume may have led to a better flowability (Caparino et al. 2012) than the spray freeze dried sample (Ishwarya et al. 2015). Parthasarathi and Anandharamakrishnan (2016) was found that the Hauser ratio of freeze-dried Vitamin E powder had 'fair to passable' flow characteristics due to their regular flaky structures which increases the frictional force whereas spray freeze dried microcapsules were represented the 'fair' flow characteristics (Table 1.2).

Flowability also depends on the drying conditions, carrier agent compositions and ratio, as seen in Table 1.2, tamarind powder with MD and GA at addition rate of 40% showed higher values of Hausner ratio and Carr index. All the tamarind powders were difficult to flow except tamarind powder with WPC at 20%, which showed medium flowing character, which may be due to its large particle size and intermediate moisture content (Bhusari et al. 2014). de Barros Fernandes et al. (2013) found that flowability values of rosemary oil powder were influenced significantly only by the temperature variable with a negative linear effect, and by the interaction of the inlet air temperature and flow rate. Furthermore, the powders obtained in this study, using wall material of hydrophobic nature, can lead to low flow. The surface composition of the powder particles plays an important role on flowability, which was also influenced by the fat loading; powders with higher levels of unencapsulated fat on the surface tended to stick together and form lumps, which impeded flow (Fitzpatrick et al. 2004).

1.3 Reconstitution Properties

Rehydration of many food powders especially produced by drying or grinding, is difficult mechanism. In terms of drying of foods, the expression of reconstitution; refers to the fact that a dried food takes its initial state of food as a result of holding or absorbing again the previously lost water in a medium containing high amounts of water (Masters 1991). Particularly, in the rehydration process, when water moves into narrow areas between small particles (particle diameter < 100 μm) with capillary forces effect, particles begin to dissolve, and a gel-like layer is formed on the surface of the particles. As a result, the progress of water between particles is prevented. This results in clusters with dry particles in the middle. In order to prevent this situation and to obtain a homogenous and uniform distribution, a strong mechanical mixing is required. Wettability, sinkability, dispersibility and solubility terms affect the reconstitution characteristics of powder. In addition, changing the drying method and the drying conditions are also leads to a change in the characteristics of rehydration. For example, the freeze drying, and osmotic dehydration methods differ from each other in the reconstitution characteristics of dried products. The most effective method for increasing the rehydration characteristics of dried food powders is agglomeration (Barletta and Barbosa-Cánovas 1993). For agglomeration of the particles, the powder product is treated with steam or hot humid air, so that condensation may form on the particle surface. Small particles around 100 μm are agglomerated to a few millimeters size for improving the wettability behavior of the particles and preventing block formation (Schubert 1987). The first step of the wetting is time control and a simple solid is wetted by the capillary force of the pores in the agglomerate (Pietsch 1999). The term of wettability is water absorption capacity on the surface of the particle. This property is widely dependent on the particle size. Small particles have a large surface area (surface area, mass ratio) and may not wet on their own. Increasing the particle size and/or agglomerated particles may reduce the agglomeration rate. The property of the particle surface area can also affect the wettability. For example; the presence of free oils on the surface of powder reduces wettability. Selective use of surfactants such as lecithin may increase the wettability of oil-containing dried powders (Barbosa-Canovas et al. 2005). Depending on the second step of dissolution; sinking or submergence is controlled by the mass of the agglomerate and it is not important that the wetting takes place easily (Pietsch 1999). Sinkability is defined as the ability of the particle to sink rapidly into the water. Sinkability mainly depends on particle size and particle density. Large and dense particles can be sink better than thin and light ones. Thanks to high air content some particles have the low density and they exhibit poor sinkability. Finally, dispersibility decreases with bulk formation and increases when the sinkability is high, whereas solubility primarily depends on the chemical composition and physical condition of the powder.

1.3.1 Wettability

Wettability is a measure of the ability of the powder product to absorb liquid under the influence of capillary forces. Generally, this depends on the particle size, density, porosity, surface tension, surface area, substances with hydrophobic character in the product and the surface activity of the particle. According to these parameters, the wettability of the powders is generally determined by the angle between the surface of the powder and the penetrating water. Therefore, the surface composition of the powder has an important function in the wetting process of the powder. The surfaces covered with hydrophobic coating materials have a wide-angle during wettability and in this situation exhibits poor wettability. However, surfaces coated with hygroscopic coating materials exhibit the opposite effect and have high wettability. Reduction of the surface tension of the water, the melting point of the oil and high temperature; helps increase the wettability of the powder products. However, the large masses block to improvement of the wettability properties of powder (Hui et al. 2008). Powders having particle size smaller than 100 μm have a tight structure and have very small gaps between the particles, which makes the wettability difficult. As a result, the particle size must be greater than 100 μm to improve wettability. In this case, the water may enter the pores of the agglomerated powder, then the particles can be wetted, and the agglomerated particles can be separated from each other rapidly to turn again to their own structure. This is important for the rapid resolution of powder (Hui et al. 2008).

1.3.2 Sinkability

After the powder product has been wetted, the gas phase surrounding each particle gradually drops into the liquid phase as the liquid begins to sink. At this stage, the powder particles also initiate the solubility process (Kelly et al. 2003). Simply sinkability is defined as the dust particles falling into the liquid phase or liquid surface (Thomas et al. 2004). Particle density and particle size is primarily factor affecting the sinkability. High density particles sink faster than those with low density. The particles have high air content have large volumes, but because of their low density, they give a low sinkability ratio (Ortega-Rivas 2009). When the particle size of the powder particles is smaller than 100 μm and the density is approximately 1.5 g/cm^3, the particles can easily sink in a solution (Schubert 1993; Hogekamp and Schubert 2003). In addition to the physical structure of a single particle, molecule types of particles are also affected sinkability. For example, agglomeration of milk powder increases depending on the result of its high total density (especially for fatty milk powder). Lactose crystallization depends on the moisture content and causes agglomeration (Aguilar and Ziegler 1994; Nijdam and Langrish 2006). In particular, effervescence and swelling properties prevent sinking of the particle significantly (Freudig et al. 1999). In the first stage of reconstitution of the powder, the density of the particle increases due to the high molecular weight of the components. Dissolution of substances such as lactose and minerals increase the density of the solution. The difference in density between the particle and the solution will gradually decrease and prevent sinking. This circumstance causes the particle to rise after the first sinking stage. In industrial and laboratory measurements, the wetting of the powder product is also considered to start sinking the particles in the solution. Considering these evaluations, terms of sinkability and wettability can be used interchangeably (Fang et al. 2007).

1.3.3 Dispersibility

Dispersibility is the ability of the flocculent or agglomerated powder product to be dispersed in the whole solution by applying low mixing conditions (Hui et al. 2008). The powder product that starts to wet and sink is rapidly separated into individual particles and then starts to dissolve in solution. The rate of dispersion is important in terms of identifying the powder product as instant (Fang et al. 2007). Researches on food powders show that dispersibility increased with particle size and decreased with percentage ratio of fine particles (≤90 μm) (Vojdani 1996). Generally, the high number of particles smaller than 125 μm creates a slurry form at the bottom of the container. Particle with high pore content and/or high density is an important factor for good dispersibility (Goalard et al. 2006).

1.3.4 Solubility

The solubility is the final stage of the reconstitution process and is a very reliable factor for the evaluation of the behavior of the powder in the liquid phase (Jouppila and Roos 1994). It is generally a key factor in determining reconstruction quality. The insoluble index is often used to evaluate solubility during reconstitution. When powder is produced from materials with high protein content, the index of insoluble material becomes an important feature, because the behavior and solubility of proteins are interdependent. The solubility of proteins is mainly affected during drying process. Because during this process, the protein molecules can be emulsified, foamed and gelled. Most importantly, the drying process causes the denaturation of proteins. The solubility of proteins varies depending on their structure, the denatured particles contained therein, and the properties of the liquid medium, such as temperature and pH. Negative changes in pH and temperature can make the normal bound structure inactivated biologically or transform into a randomly unbound structure. The unbound protein structure leads to cross-linking between two proteins, as well as to electrostatic, hydrogen bonding and disulfide interaction. Not only denaturation of proteins does not have a major effect on solubility, but also the aggregation, coagulation, and collapse of proteins can change the solubility. Researches on the solubility of milk powder, storage time and temperature increase, due to the formation of cross-protein networks on the surface of the milk powder is reported to decrease the solubility. These cross-protein networks prevent water penetration so that rehydration of powder particles can be prevented (Chen and Patel 2008).

1.3.5 The Effect of Process Method and Conditions on Reconstitution Properties

Reconstitution properties of a powder is affected by properties of raw materials, carrier materials, and moisture content, particle size, particle structure and drying methods/conditions. A quick wetting is enhanced by high porosity or large pores for big particles are desirable. Sinkability depends mainly on the particle size and density, since larger and denser particles usually sink faster than finer and lighter ones. Table 1.3 shows schematically some studies evaluating the effect of reconstitution properties from different products and different drying methods/conditions. Laokuldilok and Kanha (2015) reported that increasing the inlet air temperature increased the solubility and dispersibility of the anthocyanin powders. This result was consistent with the de Barros Fernandes et al. (2013) for microencapsulated rosemary oil powder,(Kunapornsujarit and Intipunya 2013) for longan beverage powder, (Goula and Adamopoulos 2005) for tomato, (Koç et al. 2014) for yoghurt powder, (Seth et al. 2016) for sweetened yoghurt powder, while some others the opposite (de Barros Fernandes et al. 2013; de Sousa et al. 2008). The higher inlet air temperature can remove the wetted surface of the spray dried powder better than the lower temperature and resulted in a successful reduction in lump formation. Laokuldilok and Kanha (2015) also found that freeze-dried powders had better solubility values than spray dried samples. Freeze dried powders possessed higher hygroscopicity than powders produced by spray drying because of porous structure, therefore, freeze dried powders rapidly absorb moisture from the surrounding environment, resulting in a wetted surface and introducing lump formation (Cai and Corke 2000). The solubility of the powder was increased with increased content of the maltodextrin and the DE value also affected the solubility of the spray dried powders however, DE value from DE10 to DE20 resulted in an increase in solubility, but the solubility was decreased when the DE value was increased further from 20 to 30 (Laokuldilok and Kanha 2015). At the same DE value of the maltodextrin, spray dried anthocyanin powders showed higher dispersibility than freeze dried powders (Table 1.3). Maltodextrins with higher DE values contain more sugar, which resulted in rapid moisture absorption by the powders, leading to a reduction in the glass transition temperature (Jinapong et al. 2008). Several reports attribute superior properties to other carrier agents related to water solubility, such as maltodextrin (Grabowski et al. 2006; Goula and Adamopoulos 2010; de Barros Fernandes et al. 2013).

de Barros Fernandes et al. (2013) stated that the concentration of wall material and the interaction of wall material and temperature were the variables that most influenced-on wettability due to the lower moisture content of the powders obtained under these conditions (Table 1.3). An increase in maltodextrin concentration in sumac extracts led to a significant decrease in the wettability of the powders except the sample with 25% total soluble solid concentration (Caliskan and Dirim 2013). Bhusari et al. (2014) reported that dispersibility of the tamarind powder decreased with increase in addition rate of the carrier agent. This is in agreement with the results observed by Jaya and Das (2004). Caliskan and Dirim (2013) and Koç et al. (2014) observed that the inlet/outlet temperatures had a significant effect on the wettability of powders. The wettability time of sumac extract powders increased with increasing inlet/outlet air temperatures. In conclusion the effect of carrier agent type and compositions, drying methods/conditions were changeable on the reconstitution properties of the powder. For this reason, drying conditions and carrier agent type and compositions must be optimized to achieve an effective production and good product.

Table 1.3 Effects of different drying methods/conditions on reconstitution properties

Food Powder	Drying method	Drying conditions	Wettability	Dispersibility	Solubility	Main findings	References
Rosemary oil powder	Spray	Wall material Gum arabic concentration: 10–30% (w/v) Two fluid nozzle T_{inlet}: 135–195 °C V_{feed}: 0.5–1 L/h CCRD design	In the range of 155–481 s	–	In the range of 55.75–67.75%.	These spray drying conditions led to a reduction in wetting time. The concentration of wall material influenced wettability	de Barros Fernandes et al. (2013)
Sumac extract powder	Spray	Carrier agent: Maltodextrin (DE10) Total soluble solid contents: 10–25% (w/w) Rotary atomizer Co-current T_{inlet}: 160, 180 and 200 °C T_{outlet}: 80, 90 and 100 °C $P_{atomization}$: 392 kPa	In the range of 1239–3263 s		In the range of 93.5–314.5 s	The solubility and wettability of the powders were significantly affected by both the inlet/outlet temperature, MD additions, and the temperature–MD interactions	Caliskan and Dirim (2013)
Tamarind powder	Spray	Carrier agent: Maltodextrin (20 DE), Gum arabic and whey protein concentrate concentration: 40, 50 and 60% for maltodextrin MD and GA and 10, 20 and 30% for WPC T_{inlet}: 180 °C T_{outlet}: 80 °C V_{feed}: 600 mL/h	–	In the range of 68.08–79.63%	–	Dispersibility values decreased with increase in addition rate of the carrier agent	Bhusari et al. (2014)

(continued)

Table 1.3 (continued)

Food Powder	Drying method	Drying conditions	Wettability	Dispersibility	Solubility	Main findings	References
Yoghurt powder	Spray	Rotary atomizer co-current T_{inlet}: 150–180 °C T_{outlet}: 60–90 °C T_{feed}:4–30 °C CCRD design	In the range of 307–756 s	In the range of 194–897 s	In the range of 65–72.75%	Solubility values were affected by the outlet and inlet air temperatures whereas the dispersibility and wettability values were affected by the outlet air temperatures	Koç et al. (2014)
Sweetened yoghurt powder	Spray	Carrier agent: Maltodextrin Two fluid nozzle T_{inlet}: 140–180 °C V_{feed}: 0.30–0.60 L/h $P_{atomization}$: 500–1000 kPa T_{feed}: 40 °C CCRD design	In the range of 132–378 s	In the range of 70.62–88.74%	In the range of 72–88%	Solubility of SYP was affected by all the process variables. a positive effect of inlet air temperature with wettability a negative effect of atomization pressure and inlet air temperature on dispersibility	Seth et al. (2016)
Black glutinous rice bran anthocyanins	Spray Freeze	Carrier agent: lack glutinous rice maltodextrins ((DE10, 20 and 30) T_{inlet}: 140, 160 and 180 V_{feed}: 25 mL/min −45 °C at 0.003 mBar for 48 h	–	In the range of 73.00–94.44% In the range of 73.00–81.00%	In the range of 76.23–91.79% In the range of 82.14–84.15%	Increasing the inlet air temperature increased the solubility and dispersibility of the anthocyanin powders at every DE value. The highest solubility values obtained by freeze drying. Spray dried anthocyanin powders showed higher dispersibility than freeze dried powders	Laokuldilok and Kanha (2015)

Microbial transglutaminase	Spray-freeze Freeze	Nozzle: ultrasonic, 48 kHz V_{feed}: 6.37 mL/min Main drying: 6 h at 1 mbar Secondary drying: 2 h at 0.01 mbar −80 °C for 4 h then Main drying: 6 h at 1 mbar Secondary drying: 2 h at 0.01 mbar	8.53 (±0.43) s 13.95 (±0.17) s	–	96.17 (±0.15)% 96.06 (±0.30)%	The spray freeze dried sample displayed comparatively more wettability than its freeze dried sample.	Isleroglu et al. (2018)

CCRD: central composite rotatable design, *FD*: freeze drying, *SFD*: spray freeze drying, *SD*: spray drying, *MD*: maltodextrin, *DE*: dextrose equivalent

References

Adhikari, B., Howes, T., Bhandari, B., & Troung, V. (2004). Effect of addition of maltodextrin on drying kinetics and stickiness of sugar and acid-rich foods during convective drying: Experiments and modelling. *Journal of Food Engineering, 62*(1), 53–68. https://doi.org/10.1016/s0260-8774(03)00171-7.

Aguilar, C. A., & Ziegler, G. R. (1994). Physical and microscopic characterization of dry whole milk with altered lactose content. 2. Effect of lactose crystallization. *Journal of Dairy Science, 77*(5), 1198–1204. https://doi.org/10.3168/jds.s0022-0302(94)77058-2.

Al-Kahtani, H. A., & Hassan, B. H. (1990). Spray drying of roselle (Hibiscus sabdariffa L.) extract. *Journal of Food Science, 55*(4), 1073–1076. https://doi.org/10.1111/j.1365-2621.1990.tb01601.x.

Barbosa-Canovas, G., & Juliano, P. (2005). Physical and chemical properties of food powders. In *Food science and technology* (pp. 39–71). Boca Raton: CRC Press. https://doi.org/10.1201/9781420028300.ch3.

Barbosa-Canovas, G. V., Ortega-Rivas, E., Juliano, P., & Yan, H. (2005). *Food powders: Physical properties, processing, and functionality*. New York: Kluwer Academic.

Barletta, B. J., & Barbosa-Cánovas, G. V. (1993). An attrition index to assess fines formation and particle size reduction in tapped agglomerated food powders. *Powder Technology, 77*(1), 89–93. https://doi.org/10.1016/0032-5910(93)85011-w.

Benković, M., Srečec, S., Špoljarić, I., Mršić, G., & Bauman, I. (2012). Flow properties of commonly used food powders and their mixtures. *Food and Bioprocess Technology, 6*(9), 2525–2537. https://doi.org/10.1007/s11947-012-0925-3.

Bhusari, S. N., Muzaffar, K., & Kumar, P. (2014). Effect of carrier agents on physical and microstructural properties of spray dried tamarind pulp powder. *Powder Technology, 266*, 354–364. https://doi.org/10.1016/j.powtec.2014.06.038.

Bicudo, M. O. P., Jó, J., de Oliveira, G. A., Chaimsohn, F. P., Sierakowski, M. R., de Freitas, R. A., & Ribani, R. H. (2015). Microencapsulation of juçara (Euterpe edulis M.) pulp by spray drying using different carriers and drying temperatures. *Drying Technology, 33*(2), 153–161.

Braga, M. B., dos Santos Rocha, S. C., & Hubinger, M. D. (2018). Spray-drying of milk-blackberry pulp mixture: Effect of carrier agent on the physical properties of powder, water sorption, and glass transition temperature. *Journal of Food Science, 83*(6), 1650–1659. https://doi.org/10.1111/1750-3841.14187.

Cai, Y. Z., & Corke, H. (2000). Production and properties of spray-dried amaranthus betacyanin pigments. *Journal of Food Science, 65*(7), 1248–1252. https://doi.org/10.1111/j.1365-2621.2000.tb10273.x.

Caliskan, G., & Dirim, S. N. (2013). The effects of the different drying conditions and the amounts of maltodextrin addition during spray drying of sumac extract. *Food and Bioproducts Processing, 91*(4), 539–548. https://doi.org/10.1016/j.fbp.2013.06.004.

Caparino, O. A., Tang, J., Nindo, C. I., Sablani, S. S., Powers, J. R., & Fellman, J. K. (2012). Effect of drying methods on the physical properties and microstructures of mango (Philippine 'Carabao' var.) powder. *Journal of Food Engineering, 111*(1), 135–148. https://doi.org/10.1016/j.jfoodeng.2012.01.010.

Carr, R. L. (1965). Evaluating flow properties of solids. *Chemical Engineering, 72*, 163–168. Retrieved from http://scholar.google.com/scholar?hl=en&btnG=Search&q=intitle:Evaluating+Flow+Properties+of+Solid#0.

Chan, L. C. Y., & Page, N. W. (1997). Particle fractal and load effects on internal friction in powders. *Powder Technology, 90*(3), 259–266. https://doi.org/10.1016/s0032-5910(96)03228-7.

Chang, K. S., Kim, D. W., Kim, S. S., & Jung, M. Y. (1998). Bulk flow properties of model food powder at different water activity. *International Journal of Food Properties, 1*(1), 45–55. https://doi.org/10.1080/10942919809524564.

Chegini, G. R., & Ghobadian, B. (2005). Effect of spray-drying conditions on physical properties of orange juice powder. *Drying Technology, 23*(3), 657–668. https://doi.org/10.1081/drt-200054161.

Chen, X. D. (1994). Mathematical analysis of powder discharge through longitudinal slits in a slowly rotating drum: Objective measurements of powder flowability. *Journal of Food Engineering, 21*(4), 421–437. https://doi.org/10.1016/0260-8774(94)90064-7.

Chen, X. D., & Patel, K. C. (2008). Manufacturing better quality food powders from spray drying and subsequent treatments. *Drying Technology, 26*(11), 1313–1318. https://doi.org/10.1080/07373930802330904.

de Barros Fernandes, R. V., Borges, S. V., & Botrel, D. A. (2013). Influence of spray drying operating conditions on microencapsulated rosemary essential oil properties. *Ciência e Tecnologia de Alimentos, 33*, 171–178. https://doi.org/10.1590/s0101-20612013000500025.

de Sousa, A. S., Borges, S. V., Magalhães, N. F., Ricardo, H. V., & Azevedo, A. D. (2008). Spray-dried tomato powder: Reconstitution properties and colour. *Brazilian Archives of Biology and Technology, 51*(4), 607–614. https://doi.org/10.1590/s1516-89132008000400019.

Eisen, W. B., Ferguson, B. L., German, R. M., Iacocca, R., Lee, P. W., Madan, D. et. al. (1998). *Powder metal technologies and applications*. Novelty, OH: ASM International. Retrieved from https://www.osti.gov/servlets/purl/289959.

Fang, Y., Selomulya, C., & Chen, X. D. (2007). On measurement of food powder reconstitution properties. *Drying Technology, 26*(1), 3–14. https://doi.org/10.1080/07373930701780928.

Fasina, O. O. (2007). Does a pycnometer measure the true or apparent particle density of agricultural materials? In *2007 Minneapolis, Minnesota, June 17-20, 2007*. St. Joseph, MI: American Society of Agricultural and Biological Engineers. https://doi.org/10.13031/2013.23513.

Fayed, M. E., & Otten, L. (Eds.). (1997). *Handbook of powder science and technology*. New York: Springer. https://doi.org/10.1007/978-1-4615-6373-0.

Fazaeli, M., Emam-Djomeh, Z., Kalbasi-Ashtari, A., & Omid, M. (2012). Effect of process conditions and carrier concentration for improving drying yield and other quality attributes of spray dried black mulberry (Morus nigra) juice. *International Journal of Food Engineering, 8*(1), 1. https://doi.org/10.1515/1556-3758.2023.

Fitzpatrick, J. (2005). Food powder flowability. In *Food science and technology* (pp. 247–260). Boca Raton: CRC Press. https://doi.org/10.1201/9781420028300.ch10.

Fitzpatrick, J. J. (2007). Particle properties and the design of solid food particle processing operations. *Food and Bioproducts Processing, 85*(4), 308–314. https://doi.org/10.1205/fbp07056.

Fitzpatrick, J. J., Barringer, S. A., & Iqbal, T. (2004). Flow property measurement of food powders and sensitivity of Jenike's hopper design methodology to the measured values. *Journal of Food Engineering, 61*(3), 399–405.

Freudig, B., Hogekamp, S., & Schubert, H. (1999). Dispersion of powders in liquids in a stirred vessel. *Chemical Engineering and Processing Process Intensification, 38*(4–6), 525–532. https://doi.org/10.1016/s0255-2701(99)00049-5.

Gallo, L., Llabot, J. M., Allemandi, D., Bucalá, V., & Piña, J. (2011). Influence of spray-drying operating conditions on Rhamnus purshiana (Cáscara sagrada) extract powder physical properties. *Powder Technology, 208*(1), 205–214. https://doi.org/10.1016/j.powtec.2010.12.021.

Geldart, D. (1973). Types of gas fluidization. *Powder Technology, 7*(5), 285–292. https://doi.org/10.1016/0032-5910(73)80037-3.

Goalard, C., Samimi, A., Galet, L., Dodds, J. A., & Ghadiri, M. (2006). Characterization of the dispersion behavior of powders in liquids. *Particle and Particle Systems Characterization, 23*(2), 154–158. https://doi.org/10.1002/ppsc.200601024.

Goula, A. M., & Adamopoulos, K. G. (2004). Spray drying of tomato pulp: Effect of feed concentration. *Drying Technology, 22*(10), 2309–2330. https://doi.org/10.1081/drt-200040007.

Goula, A. M., & Adamopoulos, K. G. (2005). Spray drying of tomato pulp in dehumidified air: II. The effect on powder properties. *Journal of Food Engineering, 66*(1), 35–42. https://doi.org/10.1016/j.jfoodeng.2004.02.031.

Goula, A. M., & Adamopoulos, K. G. (2010). A new technique for spray drying orange juice concentrate. *Innovative Food Science & Emerging Technologies, 11*(2), 342–351. https://doi.org/10.1016/j.ifset.2009.12.001.

Goula, A. M., Adamopoulos, K. G., & Kazakis, N. A. (2004). Influence of spray drying conditions on tomato powder properties. *Drying Technology, 22*(5), 1129–1151. https://doi.org/10.1081/drt-120038584.

Grabowski, J. A., Truong, V.-D., & Daubert, C. R. (2006). Spray-drying of amylase hydrolyzed sweetpotato puree and physicochemical properties of powder. *Journal of Food Science, 71*(5), E209–E217. https://doi.org/10.1111/j.1750-3841.2006.00036.x.

Hausner, H. H. (1967). Friction conditions in a mass of metal powder. *International Journal of Powder Metallurgy, 3*, 7–13.

Hogekamp, S., & Schubert, H. (2003). Rehydration of food powders. *Food Science and Technology International, 9*(3), 223–235. https://doi.org/10.1177/1082013203034938.

Hui, Y., Clary, C., Farid, M. M., Fasina, O. O., Noomhorm, A., & Welti-Chanes, J. (2008). Food drying science and technology: Microbiology, chemistry, applications, p. 75. Retrieved from http://www.worldcocoafoundation.org/wp-content/uploads/files_mf/phillipsmoradiseasespestsdiseasescentralamerica3.3mb.pdf

Ishwarya, S. P., & Anandharamakrishnan, C. (2015). Spray-freeze-drying approach for soluble coffee processing and its effect on quality characteristics. *Journal of Food Engineering, 149*, 171–180. https://doi.org/10.1016/j.jfoodeng.2014.10.011.

Ishwarya, S. P., Anandharamakrishnan, C., & Stapley, A. G. (2015). Spray-freeze-drying: A novel process for the drying of foods and bioproducts. *Trends in Food Science and Technology, 41*(2), 161–181.

Isleroglu, H., Turker, I., Tokatli, M., & Koc, B. (2018). Ultrasonic spray-freeze drying of partially purified microbial transglutaminase. *Food and Bioproducts Processing, 111*, 153–164. https://doi.org/10.1016/j.fbp.2018.08.003.

Jaya, S., & Das, H. (2004). Effect of maltodextrin, glycerol monostearate and tricalcium phosphate on vacuum dried mango powder properties. *Journal of Food Engineering, 63*(2), 125–134. https://doi.org/10.1016/s0260-8774(03)00135-3.

Jinapong, N., Suphantharika, M., & Jamnong, P. (2008). Production of instant soymilk powders by ultrafiltration, spray drying and fluidized bed agglomeration. *Journal of Food Engineering, 84*(2), 194–205.

Johanson, J. R. (1978). Know your material to predict and use the properties of bulk solids. *Chemical Engineer, 85*(24), 9–17.

Johanson, K. (2005). Powder flow properties. In *Food science and technology* (pp. 331–361). Boca Raton: CRC Press. https://doi.org/10.1201/9781420028300.ch13.

Jouppila, K., & Roos, Y. H. (1994). Glass transitions and crystallization in milk powders. *Journal of Dairy Science, 77*(10), 2907–2915. https://doi.org/10.3168/jds.s0022-0302(94)77231-3.

Kelly, A. L., O'Connell, J. E., & Fox, P. F. (2003). Manufacture and properties of milk powders. In *Advanced dairy chemistry—1. Proteins* (pp. 1027–1061). Boston: Springer. https://doi.org/10.1007/978-1-4419-8602-3_29.

Kelly, G. M., O'Mahony, J. A., Kelly, A. L., & O'Callaghan, D. J. (2014). Physical characteristics of spray-dried dairy powders containing different vegetable oils. *Journal of Food Engineering, 122*, 122–129. https://doi.org/10.1016/j.jfoodeng.2013.08.028.

Kha, T. C., Nguyen, M. H., & Roach, P. D. (2010). Effects of spray drying conditions on the physicochemical and antioxidant properties of the Gac (Momordica cochinchinensis) fruit aril powder. *Journal of Food Engineering, 98*(3), 385–392. https://doi.org/10.1016/j.jfoodeng.2010.01.016.

Koç, B., & Kaymak-Ertekİn, F. (2014). The effect of spray drying processing conditions on physical properties of spray dried maltodextrin. In *9th Baltic Conference on Food Science and Technology "Food for Consumer Well-Being" FOODBALT 2014*, Jelgava, Latvia, 8–9 May 2014 (pp. 243–247).

Koç, B., Sakin-Yılmazer, M., Kaymak-Ertekin, F., & Balkır, P. (2014). Physical properties of yoghurt powder produced by spray drying. *Journal of Food Science and Technology, 51*(7), 1377–1383.

Koç, M., Koç, B., Yilmazer, M. S., Ertekin, F. K., Susyal, G., & Bagdatlioglu, N. (2011). Physicochemical characterization of whole egg powder microencapsulated by spray drying. *Drying Technology, 29*(7), 780–788. https://doi.org/10.1080/07373937.2010.538820.

Koç, M., Güngör, Ö., Zungur, A., Yalçın, B., Selek, İ., Ertekin, F. K., & Ötles, S. (2015). Microencapsulation of extra virgin olive oil by spray drying: Effect of wall materials composition, process conditions, and emulsification method. *Food and Bioprocess Technology, 8*(2), 301–318. https://doi.org/10.1007/s11947-014-1404-9.

Kunapornsujarit, D., & Intipunya, P. (2013). Effect of spray drying temperature on quality of longan beverage powder. *Food and Applied Bioscience Journal, 1*, 81–89.

Kurozawa, L. E., Park, K. J., & Hubinger, M. D. (2009). Effect of carrier agents on the physicochemical properties of a spray dried chicken meat protein hydrolysate. *Journal of Food Engineering, 94*(3–4), 326–333. https://doi.org/10.1016/j.jfoodeng.2009.03.025.

Laokuldilok, T., & Kanha, N. (2015). Effects of processing conditions on powder properties of black glutinous rice (Oryza sativa L.) bran anthocyanins produced by spray drying and freeze drying. *LWT - Food Science and Technology, 64*(1), 405–411.

Liu, L. X., Marziano, I., Bentham, A. C., Litster, J. D., White, E. T., & Howes, T. (2008). Effect of particle properties on the flowability of ibuprofen powders. *International Journal of Pharmaceutics, 362*(1–2), 109–117.

Masters, K. (1991). *Spray drying handbook* (5th ed.). Harlow: Longman Scientific & Technical.

Mermelstein, N. H. (2001). Spray drying. *Food Technology, 55*(4), 92–95.

Miravet, G., Alacid, M., Obón, J. M., & Fernández-López, J. A. (2015). Spray-drying of pomegranate juice with prebiotic dietary fibre. *International Journal of Food Science and Technology, 51*(3), 633–640. https://doi.org/10.1111/ijfs.13021.

Mirhosseini, H., & Amid, B. T. (2013). Effect of different drying techniques on flowability characteristics and chemical properties of natural carbohydrate-protein Gum from durian fruit seed. *Chemistry Central Journal, 7*(1). https://doi.org/10.1186/1752-153x-7-1.

Mujumdar, A. S. (2007). Book review: Handbook of industrial drying, third edition. *Drying Technology, 25*(6), 1133–1134. https://doi.org/10.1080/07373930701399224.

Nadeem, H. Ş., Torun, M., & Özdemir, F. (2011). Spray drying of the mountain tea (Sideritis stricta) water extract by using different hydrocolloid carriers. *LWT - Food Science and Technology, 44*(7), 1626–1635. https://doi.org/10.1016/j.lwt.2011.02.009.

Nep, E. I., & Conway, B. R. (2011). Physicochemical characterization of grewia polysaccharide gum: Effect of drying method. *Carbohydrate Polymers, 84*(1), 446–453. https://doi.org/10.1016/j.carbpol.2010.12.005.

Nijdam, J. J., & Langrish, T. A. G. (2006). The effect of surface composition on the functional properties of milk powders. *Journal of Food Engineering, 77*(4), 919–925. https://doi.org/10.1016/j.jfoodeng.2005.08.020.

Ortega-Rivas, E. (2009). Bulk properties of food particulate materials: An appraisal of their characterisation and relevance in processing. *Food and Bioprocess Technology, 2*(1), 28–44.

Parthasarathi, S., & Anandharamakrishnan, C. (2016). Enhancement of oral bioavailability of vitamin E by spray-freeze drying of whey protein microcapsules. *Food and Bioproducts Processing, 100*, 469–476. https://doi.org/10.1016/j.fbp.2016.09.004.

Peleg, M. (1977). Flowability of food powders and methods for its evaluation — a review. *Journal of Food Process Engineering, 1*(4), 303–328.

Peleg, M., Mannheim, C. H., & Passy, N. (1973). Flow properties of some food powders. *Journal of Food Science, 38*(6), 959–964. https://doi.org/10.1111/j.1365-2621.1973.tb02124.x.

Peleg, M., & Moreyra, R. (1979). Effect of moisture on the stress relaxation pattern of compacted powders. *Powder Technology, 23*(2), 277–279. https://doi.org/10.1016/0032-5910(79)87018-7.

Pietsch, W. (1999). Readily engineer agglomerates with special properties from micro-and nanosized particles. *Chemical Engineering Progress, 95*(8), 67–80.

Roos, Y. H. (2003). Thermal analysis, state transitions and food quality. *Journal of Thermal Analysis and Calorimetry, 71*(1), 197–203.

Sablani, S. S., Shrestha, A. K., & Bhandari, B. R. (2008). A new method of producing date powder granules: Physicochemical characteristics of powder. *Journal of Food Engineering, 87*(3), 416–421. https://doi.org/10.1016/j.jfoodeng.2007.12.024.

Santhalakshmy, S., Bosco, S. J. D., Francis, S., & Sabeena, M. (2015). Effect of inlet temperature on physicochemical properties of spray-dried jamun fruit juice powder. *Powder Technology, 274*, 37–43. https://doi.org/10.1016/j.powtec.2015.01.016.

Schubert, H. (1987). Food particle technology. Part I: Properties of particles and particulate food systems. *Journal of Food Engineering, 6*(1), 1–32. https://doi.org/10.1016/0260-8774(87)90019-7.

Schubert, H. (1993). Instantization of powdered food products. *International Chemical Engineering, 33*(1), 28–45.

Seth, D., Mishra, H. N., & Deka, S. C. (2016). Functional and reconstitution properties of spray-dried sweetened yogurt powder as influenced by processing conditions. *International Journal of Food Properties, 20*(7), 1603–1611. https://doi.org/10.1080/10942912.2016.1214965.

Shishir, M. R. I., Taip, F. S., Aziz, N. A., & Talib, R. A. (2014). Physical properties of spray-dried pink guava (Psidium guajava) powder. *Agriculture and Agricultural Science Procedia, 2*, 74–81. Retrieved from http://linkinghub.elsevier.com/retrieve/pii/S2210784314000126.

Shrestha, A. K., Ua-arak, T., Adhikari, B. P., Howes, T., & Bhandari, B. R. (2007). Glass transition behavior of spray dried orange juice powder measured by differential scanning calorimetry (DSC) and thermal mechanical compression test (TMCT). *International Journal of Food Properties, 10*(3), 661–673. https://doi.org/10.1080/10942910601109218.

Sutton, H. M. (1976). Flow properties of powders and the role of surface character. In *Characterization of powder surfaces: With special reference to pigments and fillers* (pp. 107–158). London: Academic Press.

Teunou, E., Fitzpatrick, J. J., & Synnott, E. C. (1999). Characterization of food powder flowability. *Journal of Food Engineering, 39*(1), 31–37.

Teunou, E., Vasseur, J., & Krawczyk, M. (1995). Measurement and interpretation of bulk solids angle of repose for industrial process design. *Powder Handling and Processing, 7*(3), 203–227.

Thomas, M. E. C., Scher, J., Desobry-Banon, S., & Desobry, S. (2004). Milk powders ageing: Effect on physical and functional properties. *Critical Reviews in Food Science and Nutrition, 44*(5), 297–322. https://doi.org/10.1080/10408690490464041.

Tonon, R. V., Brabet, C., & Hubinger, M. D. (2010). Anthocyanin stability and antioxidant activity of spray-dried açai (Euterpe oleracea Mart.) juice produced with different carrier agents. *Food Research International, 43*(3), 907–914. https://doi.org/10.1016/j.foodres.2009.12.013.

Türker, İ., Koç, B., & İşleroğlu, H. (2018). Püskürtmeli-Dondurarak Kurutma İşleminin Maltodekstrinin Fiziksel Özellikleri Üzerine Etkisi. *GIDA, 43*(2), 197–210.

Vojdani, F. (1996). Solubility. In *Methods of testing protein functionality* (pp. 11–60). Boston: Springer. https://doi.org/10.1007/978-1-4613-1219-2_2.

Walton, D. E. (2000). The morphology of spray-dried particles a qualitative view. *Drying Technology, 18*(9), 1943–1986. https://doi.org/10.1080/07373930008917822.

Walton, D. E., & Mumford, C. J. (1999). The morphology of spray-dried particles: The effect of process variables upon the morphology of spray-dried particles. *Chemical Engineering Research and Design, 77*(5), 442–460.

Woodcock, C. R., & Mason, J. S. (1987). Gravity flow of bulk solids. In *Bulk solids handling* (pp. 47–83). Dordrecht: Springer. https://doi.org/10.1007/978-94-009-2635-6_2.

Yousefi, S., Emam-Djomeh, Z., & Mousavi, S. M. (2010). Effect of carrier type and spray drying on the physicochemical properties of powdered and reconstituted pomegranate juice (Punica Granatum L.). *Journal of Food Science and Technology, 48*(6), 677–684. https://doi.org/10.1007/s13197-010-0195-x.

Yue, S., Jing, W., Yubin, W., Huijuan, Z., Yue, M., Xiaoyan, Z., & Chao, Z. (2018). Inlet temperature affects spray drying quality of watermelon powder. *Czech Journal of Food Sciences, 36*(4), 316–323. https://doi.org/10.17221/406/2017-cjfs.

Zou, R. P., & Yu, A. B. (1996). Evaluation of the packing characteristics of mono-sized non-spherical particles. *Powder Technology, 88*(1), 71–79. https://doi.org/10.1016/0032-5910(96)03106-3.

Chapter 2
Food Powders Particle Properties

Ulaş Baysan, Mehmet Koç, and Banu Koç

2.1 Particle Shape

Single particle characteristics are crucial for product properties. The size, shape, surface, density and hardness of a single particle are the main properties for characterization of powder products (Davies 1984). Particle shape is the one of the most important properties due to classify the particles. The size of simple shaped particles such as spheres and cylinders is determined by one or several dimension. A number of dimensions are used in defining the size and shape of the irregularly shaped particles. The numerous of shape factors such as sphericity, width and convexity are used in the determination of the shapes of food powders (Murrieta-Pazos et al. 2012; Saad et al. 2011). Food powders' origin is generally organic material and their shapes are showed differences one other because their chemical structure are more complex than inorganic powders produced by food industry. Extreme irregular shapes (generally food materials obtained by milling), shapes close to the global structure (starch), or crystalline forms (salt and sugar) may be faced at food powders structure (Dhanalakshmi et al. 2011). The particle shapes can be classified as acicular (needle-like pointed), angular, crystallized, branched, fibrous, pulsed, granular (in the form of symmetrical sized particles), irregular, modular (rounded irregular shape) and spherical shapes (Ermiş 2015). The particle shape is one of the most beneficial factor for determining particle behavior and structure. However, these simple categorizations may not be enough to compare the particle size or investigate the equations added variable of the particle shapes. Thus, just looking at

U. Baysan · M. Koç (✉)
Aydın Adnan Menderes University, Faculty of Engineering, Department of Food Engineering, Aydın, Turkey
e-mail: mehmetkoc@adu.edu.tr

B. Koç
Gaziantep University, Fine Arts, Gastronomy and Culinary Arts Department, Gaziantep, Turkey

E. Ermiş (ed.), *Food Powders Properties and Characterization*, Food Engineering Series, https://doi.org/10.1007/978-3-030-48908-3_2

the particle shapes may not be useful without the quantitative approach. The particle size is one of the preferred in these quantitative approaches.

The particle size plays a vital role for powder characterization, categorization and classification. A particle material is defined as powder when its median size is less than 1 mm (half number of material is larger than the median size, the other half is smaller). Although the powder products are expressed as "fine" and "coarse", the most of powder foods are considered to be fine-grained (Koç et al. 2011).

Industrial inorganic powder products are usually obtained by reducing the size of a hard material (milling). The resulting particles are multi-face (4–7), edged and cornered. For this reason, "diameter" expression is used as the characteristic linear dimension in the determination of particle size (McCabe et al. 2005). Due to their organic chemical structure, the particulate food compositions are quite complex compared to inorganic industrial products. As a result, when examined in terms of shape, powdered foods can vary from highly amorphous (spices) to highly curative (starch) structure. Equivalent spherical diameter, equivalent circular diameter and statistical diameter methods are used to determine the size and diameter of irregular particles. It is aimed to determine the diameter of the sphere with the same properties as the particle at the equivalent spherical diameter method. Moreover, the diameter of the circle with the same environment or visible area is determined in the equivalent circular diameter method. According to the statistical diameter method, the particle size is usually determined by microscope observation.

An irregular powder particle can be examined by various characterization techniques. The different parameters of the particle are evaluated by these different techniques. Each of these different evaluations is used for different purposes. Therefore, the techniques using should be chosen carefully for different particle and situation. For example, the Stoke diameter of powder is used at investigation of the sedimentation of grain. The Stokes diameter is defined as "the size of a smooth sphere of the same density as the powder grain that has the same settling speed as the fine particle at low Reynolds number in a viscous fluid." (Fayed and Otten 2013). In suspension having particles densely, the particles move downward when there is an upward movement of the fluid. Thus, the particles mix with one another and affect each other's movements. The Stokes law determines the velocities proportional to the square of the particle diameters (Wang and Langrish 2009). Moreover, volumetric diameter of powder products ($D_{4,3}$) helps to storage of food particles, packaging, transportation cost, and process planning. The size of the packaging material and the storage area is also determined by the volumetric diameter. It is determined that the volumetric diameter had direct effect on the bulk properties (Koç et al. 2011). The surface diameter of particle ($D_{3,2}$) is utilized at situation where the adsorption mechanism is active. In addition, the usage areas of the products such as starch and flour can be determined by the surface diameters of the particles. However, the determination of the different particle diameters mentioned above is not generally preferred at food industry due to the long time, an increase in workload and equipment cost. As a result of its practicality, fastness and low cost of equipment, sieve diameter is utilized to determine particle size at food industry.

Determination of the relevant characteristic particle size is faced as a challenge since particle shapes are far from sphericity. Therefore, it is very difficult to accurately measure the particle size and investigation of the particle size alone is not accurate. The idea of determining the general distribution of particles occurs at this point.

A single particle size may not represent the entire population. Powder products can have very different sizes even if they have the same chemical composition. Particle distribution refers to the distribution of particle size. Particle distribution is a powder product characteristic that is frequently used in a physical, mechanical or chemical process to determine the behavior and physical properties of the material. Food powders are usually sold in fine powder form (Schubert 1987). Although there are many different types of measurement techniques to determine particle distribution, there are four commonly used methods. These are sieve analysis, microscope technique, sedimentation, and laser light scanning technique.

The principle of sieve analysis is geometric similarity of particle. It is a useful, easy, inexpensive, and highly reproducible method to determine the particle distribution. Therefore, the particle distribution of the powder product is determined by determining the mass of particles in the same size range (Shenoy et al. 2015).

In the microscope technique, the particle distribution is determined by directly measuring and counting the size of the particles under an optical microscope. The information about the shape and morphology of the powder product can be obtained when electron microscope is also used to measure and count particle directly as optical microscope. The most widely used and preferred electron microscope is scanning electron microscope (SEM) (Addo et al. 2019; Einhorn-Stoll 2018; Karasu and Ermis 2019; Camargo Novaes et al. 2019).

Particle behavior in a suspension and the Stokes diameter of particle is defined with sedimentation technique. Gravimetric and centrifugal sedimentation techniques are used to determine the particle size distribution. Sedimentation analysis has some challenge. The most remarkable in these challenges are that the particle, which does not resemble the spheres, does not cause subside in vertical direction so particle distribution is showed deviations (Wang and Langrish 2009).

Laser distribution is the most commonly used method in order to determine particle size distribution. In this method the particles must be dispersed in a liquid or gas phase. It is called "the wet measurement" if the particle dispersed in a liquid phase, while "the dry measurement" is defined when particles are dispersed in a gas phase. The principle of this method is to determine the distribution angle that the light emitted from the laser source forms when the particles are in the liquid or gas dispersion. Laser dispersion angle increases with reduction of particle size. This method is the most preferred and applied method in the literature (Fitzpatrick et al. 2016; Yohannes et al. 2018).

The particle size distribution graphs are desired like Fig. 2.1a. However, in practice this is not generally possible, but it is desirable that this distribution exhibits a homogeneous behavior as shown in Fig. 2.1b. A very wide range of dispersion structure shows that there are particles in very different particle sizes. In other words, the mod value of the particle distribution is lower than homogenous

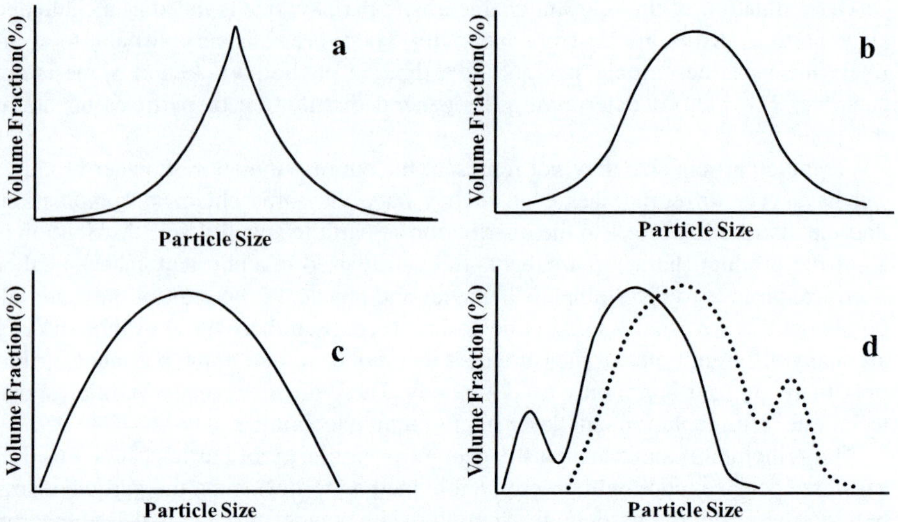

Fig. 2.1 Particle size distribution. (**a**) desired distribution (peak), (**b**) homogeneous distribution, (**c**) wide range distribution (dispersion), (**d**) bimodal distribution

distribution graph (Fig. 2.1c). The bimodal distribution on the right or left shows a distribution far from undesired homogeneity (Fig. 2.1d). The reason of this graph encountered is that the particle distribution on the right or on the left is differentiated in different particle size.

2.2 Particle Density

The particle density is determined by dividing the total mass of particle by its total volume. It is directly related to macrostructure and particle size of the powder product (Anon 2006). In addition, the density of the particles directly affects the reconstitution properties of powdered foods. High particle density and low amount of trapped air (air that has remained closed in the particle) increase the rate of penetration of powder products in liquid media (Carić and Milanović 2002). Particularly, spray-drying process conditions (air inlet temperature; air outlet temperature and atomization) used in the production of powder products from liquid foods is also effective on particle density. In addition, the dry matter content of the liquid food to be dried in the spray dryer also changes the particle density of the powder product.

Different particle densities can be defined as the true, apparent and effective (or aerodynamic) particle density considering the total volume. The true particle density is determined by proportioning the particle mass to the particle volume apart from open and closed pores. It is the particle density of organic or inorganic pure chemical products which quoted in the reference books. The apparent particle density is determined by proportioning the particle mass to the particle volume excluding the open pores only. It is measured using liquid or air pycnometer. The effective (aerodynamic) particle density is determined by dividing the particle mass to the particle volume with the open and closed pores. In this method, the particles are surrounded by a gas. This density is very important in fluidized bed method. It should be noted that the three different particle densities are not exactly defined the same as the bulk density. The void between the particles is also taken into account when measuring the bulk density.

Liquid and gas pycnometer are commonly used in the industry to determine particle density. Depending on the volume of the liquid pycnometer, the particle density of the thin or bulky materials can be determined. Larger and calibrated containers are needed for bulky-grained products, whereas a 50 mL volume pycnometer is sufficient for fine grain powder products. The liquid to be used should not react with the powder product and the powder should not dissolve in this liquid. In the gas pycnometer measurements, air or any gas (such as helium) can be used as fluid. Several food powders density determined with gas pycnometer (Börjesson et al. 2016; Pinto et al. 2018; Turchiuli et al. 2014). However, it is difficult to analyze high-porosity freeze-dried materials with this method (Stange et al. 2013).

Although determination of particle density helps us comment and explain many issues, sole particle density is not sufficient to explain all the chemical and physical properties. Therefore, the morphology of the particles should also be examined to characterize the particle.

2.3 Particle Morphology

The particle morphology is more than its shape and it provides information on the surface characteristics and structure of the particle. Numerical visuals of particles can be produced with the help of microscopes that are connected with computers. Thanks to these 2-dimensional images of particle outlines, particle morphology can be quantitatively analyzed. Although optical microscope is also used to evaluate the morphology of powders, electron microscope is more appropriate to characterize shape and morphology of particles due to its range and resolution. Because the image provided from optical microscope gives the data about only average size parameters (diameter, area, perimeter, etc.) which are inadequate for characterizing the morphology of the particles (Mikli et al. 2001). That's why the particle morphology of food powders was generally studied with scanning electron microscope (SEM) in the literature. Scanning electron microscope provides the opportunity to analyze the material to be examined with electron beam which is formed in vacuum environment and thinned with electromagnetic lenses in the same environment to create high resolution picture. The images created in the scanning electron microscope are formed by counting the emitted radiation or the reflected electrons from the interaction of the electron beam with the material. These are secondary electron image, backscattered electrons, characteristic x-rays, Auger electrons, etc. Organic and inorganic materials morphology can be analyzed by SEM. The electrons sent on the sample during the SEM examination give the sample a load. This disrupts the quality and resolution of the image. For this reason, the samples to be examined in the SEM must be conductive so that they do not hold the load on it. In order for the surface to be displayed perfectly, it must be covered with a substance that will reflect the electrons. Non-conductive samples are usually made conductive with gold-palladium, platinum or aluminum. Conductive metals do not need to be coated. The coating process is carried out in different thicknesses in special equipment and better viewing can be achieved. The air in the coating medium is replaced by argon gas and the coating is carried out by the gold-palladium which passes through the plasma phase. The most of bulk and particle properties are associated to particle morphology (Walton and Mumford 1999). The different particle morphologies that occur depending on the temperature give information about the density, size, moisture content and durability of the particles (Alamilla-Beltran et al. 2005). Walton (2000) also explained that the morphology of final particle was related to process conditions.

2.4 The Effect of Process Method and Conditions on Particle Properties

The importance of the process techniques and conditions used in order to manufacture the food powder is increasing with the increasing demand for powder products since the physical and chemical properties of the powder product are affected by the application technique and conditions. The most used methods are the drying and/or grinding process to obtain food powder. Although the powder product can be obtained by the drying-milling and milling-drying processes one after the other, it is also obtained by applying the drying process alone.

Grinding process is defined as “the large solid unit masses are reduced into the smaller unit masses”. The grinding process is widely used in the food industry to transform from the large solid units into coarse particles or fine particles. Although grinding process is also called size reduction, the process of solid size unit reduced by the mechanical forces is called milling. The ideal particle distribution should be in equality considering particle size and shape. In this case, size separation and classification are one of the points to consider for the size reduction. The different grinding techniques are selected and applied for different raw material, product and purposes since the grinding techniques or conditions affects directly on the particle properties (Patel et al. 2014). The most preferred and used grinding types in the food industry are described below.

One or more cylinders made in heavy steel are rotated to each other towards since it is provided that feed samples are captured in *disc mills***.** Two or more cylinders made in heavy steel are rotated to each other towards since it is provided that feed samples are captured. The particle size reduction results from effect of compressive force in this system. There are two different types of disc mills such as single-disc mills and double-disc mills. Most of food samples that are not hard material like flour, soybeans, and starch can be effectively reduced size particle. Bayram and Öner (2005) studied on the comparison of using the disc and hammer mills for Bulgur particles. Bulgur particles having the smooth and regular shaped was produced by disc mills in contrast to hammer mills. It is found that bulgur particle has a homogeneous distribution.

The impact and attrition forces are applied in *ball mill* which consists of hollow cylinder mounted on the metallic frame and horizontal steel cylinder which is filled 30–50% with steel balls. The small balls are preferred at low rotational speeds, while the larger balls are chosen at high rotational speeds. Fine particle can be obtained with using the balls mills such as food colorants. Vogel et al. (2018) claimed that a longer crushing time affected the microstructure of the flour particles and the crushing time can be determined carefully.

Hammer mills works with the mechanism of impact between moving hammers on rotor and the powder material. A high-speed rotor inside the chamber is equipped with hammers. Hammer mill is generally used for crystal-like and fibrous materials including spices and sugar. Mani et al. (2006) reported that the larger the sieve openings of hammer mill resulted in the lower the particle densities of samples.

Stress and attrition are main operating mechanism of *roller mills*. Stress is provided from rotating rollers or heavy wheels. Two or more steel rollers turn towards each other and the food materials are fed through the gap between the rollers. This distance between two rollers should be adjustable for different materials considering initial and final particle size, physical pollution (metals and stones) and durability. Wheat grinding is generally carried out with roller mill. Moreover, an increase in the shear and pressure forces brought about an increase in particle size of flour grinded by roller mill (Opáth 2014).

Following the sample moisture content reduction by most of the drying methods, powder product can be obtained directly (exception of drum, fluidized bed and spray drying). However, the powder can be obtained by grinding after samples dried. In other words, drying and grinding are carried out as a combined process. Although the properties of samples, the drying methods and conditions are effective on the particle properties, in this case the most effective process on the particle properties is the grinding step. In the grinding stage, milling types are effective on particle properties as described above.

Drying is one of the effective methods to manufacture the powder. Drying is a process of evaporating the water in the product by applying heat (Lewicki 2006). Advantage of the drying process is to reduce packaging and transportation costs due to the low volume and weight of the dried products, besides, dry products have a wider range of applications and longer shelf life compared to wet products. Drying is a combined mechanism of heat and mass transfer (Hernandez et al. 2000). During drying, the heat required for the evaporation of the water is transferred to the product, where mass transfer takes place in the form of water vapor from the product depending on the heat transmitted. Heat transfer mechanism is classified into three groups; convection, conduction and radiation. Therefore, drying processes can be classified according to all these heat transfer mechanisms.

The food powders can be produced directly by spray, drum and fluidized bed dryer. Although, powder product can be obtained with fluidized bed, drum or spray drying methods, while combined drying of spray and fluidized bed are used in food industry for the different usage areas and the large particle size desired since the particle produced from the spray dryer is defined the fine powder.

The drum dryers' main equipment is one or two hollow metal cylinders. This cylinder or these cylinders rotate on horizontal axis and are heated by steam, hot water or other heat sources. Thus, the drum dryer provide to the indirect heat transfer on the samples (Gavrielidou et al. 2002). The process variables of drum dryer are feed samples specific properties and concentration, vapor pressure, valse rotational speed and the gap between the two valse (it is only as this variable for drum dryer with two cylinders) (Kostoglou and Karapantsios 2003; Valous et al. 2002). The feed material composition and concentration, the water binding capacity, the drying rate, the moisture content, the rheological and thermal properties, as well as the thickness on the surface of the valse have a considerable effect on the powder product properties especially particle properties (Pua et al. 2010).

Caparino et al. (2012) studied on mango powder produced by drum dryer. It was observed that the mango powder had irregular morphology with sharp edges, compact microstructure and huge of indendation surface due to the crushing of the powder by the drum dryer. It is determined that one side of powder dried by drum dryer, which is the contact with valse surface, form smooth surface. However, the other side of powder surface is formed corrugation and indentation (Anastasiades et al. 2002). Desobry et al. (1997) compared the properties of powder products obtained from two different dryers which are drum, and spray dryer. It was observed that morphological and dimensional properties of particles were significantly differences. Samples dried by spray dryer were the spherical shape as expected due to indicating the least collapse structure. However, the powder produced from the drum dryer had a complex structure since the particles coagulated each other.

In the fluidized bed drying, small solid particles come into contact with air and these solid particles are activated. As a result of this step, it is ensured that the solid particles are suspended in the bed so the powder product is obtained by fluidized bed. Application of the fluidized bed dryer effects on the physical properties and structure of the powder such as particle size, density, shape (Szulc and Lenart 2013) as well as the behavior of reconstitution (Knight 2001). Calban and Ersahan (2003) claimed that an increase in air flow rate and drying temperature brought about the drying rate. Increasing the drying rate meant rapid mass transfer. Thus, the particle size of powder decreased and the surface area of powder is increased. Change of particle structure caused the change of particle and bulk properties.

The principle of the spray-drying process is that atomization of liquid material into a drying chamber, the encounter of droplets in the chamber with a heated gas stream to drying, separation of powder products from the air stream (Liu et al. 2016). The process conditions of spray drying are inlet and outlet temperatures, feed rate, drying gas flow rate and atomization parameters (Gharsallaoui et al. 2007). The particle formation process and particle properties were affected by these various drying process variables. Furthermore, the feed characteristic affects the particle properties directly in spray drying method (Paudel et al. 2013).

The physical properties of products are affected by inlet drying air temperature (Jafari et al. 2017). The higher inlet air temperature brings about acceleration of the heat transfer mechanism. The crust formation on the surface of particle, which is difficult in water diffusion, result from the high heat transfer (Largo Ávila et al. 2015; Moghaddam et al. 2017). At high drying temperatures depending on pressure differences between the relative humidity of drying air and the water in droplets, the bound water in the droplets will be removed by capillary forces so, the outer shell of particles gets damaged. As a result, cracking occurs on the particle surface and for this reason, the morphology of the particles, which is highly influenced by use of high drying temperatures. Moreover, the structure of a powder dried at a higher air temperature become more porous (de Souza et al. 2015). It is generally known that lower moisture content of the product usually depend on use of higher inlet air temperature (Largo Ávila et al. 2015).

The low moisture content obtained at use of high inlet air temperature due to the high drying rate and it depends not only on the temperature but also on the dry matter content of the feed, the spray droplet size, and the feed rate. However, it is generally known that lower moisture content of the product usually depend on use of higher inlet air temperature (Largo Ávila et al. 2015). The lower moisture content links to the higher particle density (Baysan et al. 2019).

The droplet size, drying kinetics and moisture content of product are affected by the feed rate so this parameter affects physical properties of the powder. The high feed flow rate have a negative effect on the moisture content of powder (Khalilian Movahhed and Mohebbi 2016). The size of the droplets increases and the contact time between droplets and drying air decreases at the higher flow rate. The drying process is completed rapidly before the drying of droplets is uncompleted because of the rapid contact time in drying chamber. Moreover, drying occurs in a larger surface area that affects the heat and mass transfer negatively. Therefore, the high particle size, the low particle density and undesirable particle structure formation are observed at high feed rate (Can Karaca et al. 2015). Moreover, the feed rate affects the outlet temperature directly since both of these factors are dependent variables each other.

The high atomization speed or pressure provides larger drying surface area due to the smaller droplet at constant feed viscosity (Largo Ávila et al. 2015). The large drying area means the effective heat and mass transfer, short diffusion path. The larger drying surface area influences powder properties positively. In illustration, Gallo et al. (2011) was claimed that atomization pressure or speed impacts significantly on particle size of powder, since the particle size of the powders, which is related to the droplet size. An increase in atomization pressure or speed cause the smaller droplet and lower size of powders (Chegini and Ghobadian 2005). Furthermore, the crust formation on the particle surface is blocked by high atomization pressure or speed (Balesdent et al. 2000). Thus, the particle can be uniform structure and shrinkage.

Other important variable of spray dryer is the viscosity, density and temperature of feed samples since these factors is effect on the powder properties directly. The feed atomization of high concentration leads to the large droplets since powder is came out the high particle density (Abiad et al. 2014).

2.5 The Effect of Particle Properties on Physicochemical Properties of Food Powder

Particle properties of powder are responsible for changes in porosity, flowability, reconstitution properties, chemical stability, color, as well as functional properties of food powders. Especially encapsulation stability of food powders in terms of oxidation stability and flavor retention highly depends on particle properties. Initial feed concentrations, drying methods, emulsion formation, emulsion preparation conditions such as homogenization type, rate and time, and wall material ratio have a big impact on the particle microstructure that is a key characteristic for many functional and fundamental properties of the final powder.

Particle structure, size, shape and interaction between particles is affected by forces such as van der Waals, capillary, electrostatic, and mechanical interlocking. Molecular adsorption can be due to these forces which are undoubtedly underpin the mechanisms of particle adhesion and cohesion that leads to particle aggregation (Hickey and Giovagnoli 2018). Cohesion and adhesion can be described in terms of stickiness, caking, aggregation and crystallization which can be correlated to the water activity of the system through the glass-transition temperature (Boonyai et al. 2004). The glass-transition temperature explains the process of structure formation such as shrinkage and collapse during drying and other related process. According to this concept, if the process takes place at a temperature below the glass transition temperature, a more porous (negligible collapse) product can be obtained. However, if the difference between the glass transition temperature and the process temperature rises, the porosity disappears and the collapse occurs clearly. In addition, micro-structure is dependent on drying technique, feed concentration, the degree of feed aeration, the interaction between drying agent/wall material and the surface.

The surface morphology and internal micro-structure of the powder is the link between processing and flavor retention, prevent oil oxidation and color (Chen and Özkan 2007). Foods are a system of multi-components, protein and carbohydrates in this structure are very important because of their different functional and textural properties (Molina-Sabio and Rodríguez-Reinoso 2004). Flavor retention of a food product after drying can also vary depending on how the drying is conducted, as it can create a porous structure that limits certain flavor compounds from escaping out of the structure. On the contrary, powder having bioactive compounds that could be affected by deterioration reactions depending on exposed area (i.e. oxidation). Furthermore, the reconstitution performance of the porous structure of powder allows the solvent to penetrate and disperse throughout (Pietsch 2008). If the surface of spherical shaped particle does not composed breaks or cracks, the permeability of gases reduced and core material is better protected (Fu and Wang 2011). Botrel et al. (2014) reported that the lower porosity in microcapsules and the fewer spaces among particles brings about the higher oxidation stability.

Although the effects of particle properties of final powder on encapsulation efficiency of aroma components and oils studied by numerous studies, the effect of particle properties on encapsulation efficiency is not clear. The particle properties of encapsulated aroma powders was found significantly important for aroma retention by (Janiszewska and Witrowa-Rajchert 2009). The encapsulation efficiency of aroma materials decreased in an increasing in particle size and a decreasing in particle density. However, Rulkens and Thijssen (2007) and Reineccius (1989) reported that the larger particle size resulted in higher encapsulation efficiency for aroma materials.

Some of studies in literature also supported these results (Blakebrough 1973; Fang 2004; Jafari et al. 2008). On the other hand, Ré (1998), Finney et al. (2002), and Soottitantawat et al. (2005) found that particle properties had no effect on encapsulation efficiency of aroma materials. In another study, the intermediate sized particles were superior to large sized particles in terms of oil retention ((Chang et al. 1988). Although the effect of particle size on encapsulation efficiency is not clear, it is necessary to produce large particles to improve rehydration characteristics of food powders. With the reduction in particle size, the dispersibility of the powder product deteriorates especially in cold water and the particles remain as a heap on the liquid surface (Jafari et al. 2010). As mentioned before, the drying process conditions determine the particle properties. For example, a higher drying temperature in spray drying allows to form fast capsules wall, thus the capsules with a larger particle size can be produced (Alamilla-Beltran et al. 2005; Finney et al. 2002; Janiszewska and Witrowa-Rajchert 2009; Walton and Mumford 1999). Jafari et al. (2010) approached from a different perspective to the effect of particle size on encapsulation efficiency. They reported that the encapsulation efficiency of aromatic oils altered with particle size within the same bulk. The larger particles (>63 mm) were superior to the smaller particles (<38 mm) considering aroma retention.

Besides the effect of particle properties of food powders on their chemical properties, the bulk properties of food powders are also highly changed with particle properties in terms of particle size and distribution, morphology and density. Powders with uniform surface and low particle size had higher bulk density (Baysan et al. 2019). Koç et al. (2015)) also found that the higher particle size resulted in the higher bulk density in olive oil powder. Low bulk density of food powder was not appropriate, due to the possibility of oxidation (Samaram et al. 2014). The flowability of food powders is also correlated with particle size. The fine particle size of food powders leads worse flowability (Koç et al. 2011). Particle properties of powder has also a vital effect on the solubility and wettability. Baysan et al. (2019) demonstrated that an increase in particle density provided a better solubility and shorter the wettability time. The color of food powders is also affected from particle properties. The color of food powders also depends on particle properties.

Fernández-López et al. (2002) found that the particle size of paprika powder determines the paprika color perception. The color of the paprika samples was perceived as lighter with the increase in the particle size, whereas the color was perceived as darker when the size was small. Conversely, Ishwarya and Anandharamakrishnan (2015) found that freeze dried coffee had a broader particle size distribution against spray and spray-freeze dried samples and hence the greater color difference. Spray ($D_{4,3}$ = 5.4 μm) and spray-freeze dried ($D_{4,3}$ = 8.7 μm) coffee samples shown a greater L value than the freeze dried ($D_{4,3}$ = 636.8 μm) sample. Sharma et al. (2013) also observed similar results in powder products.

References

Abiad, M. G., Campanella, O. H., & Carvajal, M. T. (2014). Effect of spray drying conditions on the physicochemical properties and enthalpy relaxation of α-lactose. *International Journal of Food Properties, 17*(6), 1303–1316. https://doi.org/10.1080/10942912.2012.710287.

Addo, K. A., Bi, J., Chen, Q., Wu, X., Zhou, M., Lyu, J., & Song, J. (2019). Understanding the caking behavior of amorphous jujube powder by powder rheometer. *LWT, 101*, 483–490. https://doi.org/10.1016/j.lwt.2018.11.059.

Alamilla-Beltrán, L., Chanona-Pérez, J. J., Jiménez-Aparicio, A. R., & Gutiérrez-López, G. F. (2005). Description of morphological changes of particles along spray drying. *Journal of Food Engineering, 67*(1–2), 179–184.

Anastasiades, A., Thanou, S., Loulis, D., Stapatoris, A., & Karapantsios, T. D. (2002). Rheological and physical characterization of pregelatinized maize starches. *Journal of Food Engineering, 52*(1), 57–66. https://doi.org/10.1016/s0260-8774(01)00086-3.

Anon. (2006). *Powder technology handbook*. Boca Raton: CRC Press. https://doi.org/10.1201/9781439831885.

Balesdent, J., Chenu, C., & Balabane, M. (2000). Relationship of soil organic matter dynamics to physical protection and tillage. *Soil and Tillage Research, 53*(3–4), 215–230.

Bayram, M., & Öner, M. D. (2005). Stone, disc and hammer milling of bulgur. *Journal of Cereal Science, 41*(3), 291–296. https://doi.org/10.1016/j.jcs.2004.12.004.

Baysan, U., Elmas, F., & Koç, M. (2019). The effect of spray drying conditions on physicochemical properties of encapsulated propolis powder. *Journal of Food Process Engineering, 42*(4), e13024. https://doi.org/10.1111/jfpe.13024.

Blakebrough, N. (1973). Fundamentals of fermenter design. In *Microbial engineering* (pp. 305–315). London: Butterworths. https://doi.org/10.1016/b978-0-408-70549-3.50006-8.

Boonyai, P., Bhandari, B., & Howes, T. (2004). Stickiness measurement techniques for food powders: A review. *Powder Technology, 145*(1), 34–46. https://doi.org/10.1016/j.powtec.2004.04.039.

Börjesson, E., Innings, F., Trägårdh, C., Bergenståhl, B., & Paulsson, M. (2016). Permeability of powder beds formed from spray dried dairy powders in relation to morphology data. *Powder Technology, 298*, 9–20.

Botrel, D. A., de Barros Fernandes, R. V., Borges, S. V., & Yoshida, M. I. (2014). Influence of wall matrix systems on the properties of spray-dried microparticles containing fish oil. *Food Research International, 62*, 344–352. https://doi.org/10.1016/j.foodres.2014.02.003.

Calban, T., & Ersahan, H. (2003). Drying of a Turkish lignite in a batch fluidized bed. *Energy Sources, 25*(12), 1129–1135. https://doi.org/10.1080/00908310390233568.

Camargo Novaes, S. S., Hellmeister Dantas, F. B., Alvim, I. D., Rauen de Oliveira Miguel, A. M., Vissotto, F. Z., & Vercelino Alves, R. M. (2019). Experimental method to obtain a uniform food powder mixture of omega-3 microcapsules and whole milk powder. *LWT, 102*, 372–378. https://doi.org/10.1016/j.lwt.2018.12.037.

Can Karaca, A., Guzel, O., & Ak, M. M. (2015). Effects of processing conditions and formulation on spray drying of sour cherry juice concentrate. *Journal of the Science of Food and Agriculture, 96*(2), 449–455. https://doi.org/10.1002/jsfa.7110.

Caparino, O. A., Tang, J., Nindo, C. I., Sablani, S. S., Powers, J. R., & Fellman, J. K. (2012). Effect of drying methods on the physical properties and microstructures of mango (Philippine 'Carabao'var.) powder. *Journal of Food Engineering, 111*(1), 135–148.

Carić, M., & Milanović, S. (2002). Milk powders. Physical and functional properties of milk powders. In *Encyclopedia of dairy sciences*. Oxford: Elsevier.

Chang, H. R., Rudareanu, F. C., & Pechère, J.-C. (1988). Activity of A-56268 (TE-031), a new macrolide, against *Toxoplasma gondii* in mice. *The Journal of Antimicrobial Chemotherapy, 22*(3), 359–361. https://doi.org/10.1093/jac/22.3.359.

Chegini, G. R., & Ghobadian, B. (2005). Effect of spray-drying conditions on physical properties of orange juice powder. *Drying Technology, 23*(3), 657–668. https://doi.org/10.1081/drt-200054161.

Chen, X. D., & Özkan, N. (2007). Stickiness, functionality, and microstructure of food powders. *Drying Technology, 25*(6), 959–969. https://doi.org/10.1080/07373930701397400.

Davies, R. (1984). *Particle size measurement: Experimental techniques*. New York: Van Nostrand Reinhold.

de Souza, V. B., Thomazini, M., Balieiro, J. C. d. C., & Fávaro-Trindade, C. S. (2015). Effect of spray drying on the physicochemical properties and color stability of the powdered pigment obtained from vinification byproducts of the Bordo grape (Vitis labrusca). *Food and Bioproducts Processing, 93*, 39–50. https://doi.org/10.1016/j.fbp.2013.11.001.

Desobry, S. A., Netto, F. M., & Labuza, T. P. (1997). Comparison of spray-drying, drum-drying and freeze-drying for β-carotene encapsulation and preservation. *Journal of Food Science, 62*(6), 1158–1162. https://doi.org/10.1111/j.1365-2621.1997.tb12235.x.

Dhanalakshmi, K., Ghosal, S., & Bhattacharya, S. (2011). Agglomeration of food powder and applications. *Critical Reviews in Food Science and Nutrition, 51*, 432.

Einhorn-Stoll, U. (2018). Pectin-water interactions in foods–from powder to gel. *Food Hydrocolloids, 78*, 109–119.

Ermiş, E. (2015). Gida Tozlari: Özellikleri Ve Karakterizasyonu. İçindekiler/Content. *GIDA, 40*, 287.

Fang, F. C. (2004). Antimicrobial reactive oxygen and nitrogen species: Concepts and controversies. *Nature Reviews. Microbiology, 2*(10), 820–832. https://doi.org/10.1038/nrmicro1004.

Fayed, M., & Otten, L. (2013). *Handbook of powder science and technology*. New York: Springer.

Fernández-López, J., Pérez-Alvarez, J. A., Sayas-Barberá, E., & López-Santoveña, F. (2002). Effect of paprika (Capsicum annum) on color of Spanish-type sausages during the resting stage. *Journal of Food Science, 67*(6), 2410–2414.

Finney, J., Buffo, R., & Reineccius, G. A. (2002). Effects of type of atomization and processing temperatures on the physical properties and stability of spray-dried flavors. *Journal of Food Science, 67*(3), 1108–1114. https://doi.org/10.1111/j.1365-2621.2002.tb09461.x.

Fitzpatrick, J. J., van Lauwe, A., Coursol, M., O'Brien, A., Fitzpatrick, K. L., Ji, J., & Miao, S. (2016). Investigation of the rehydration behaviour of food powders by comparing the behaviour of twelve powders with different properties. *Powder Technology, 297*, 340–348. Retrieved from https://linkinghub.elsevier.com/retrieve/pii/S0032591016302042.

Fu, F., & Wang, Q. (2011). Removal of heavy metal ions from wastewaters: A review. *Journal of Environmental Management, 92*(3), 407–418. https://doi.org/10.1016/j.jenvman.2010.11.011.

Gallo, L., Llabot, J. M., Allemandi, D., Bucalá, V., & Piña, J. (2011). Influence of spray-drying operating conditions on Rhamnus purshiana (Cáscara sagrada) extract powder physical properties. *Powder Technology, 208*(1), 205–214. https://doi.org/10.1016/j.powtec.2010.12.021.

Gavrielidou, M. A., Valous, N. A., Karapantsios, T. D., & Raphaelides, S. N. (2002). Heat transport to a starch slurry gelatinizing between the drums of a double drum dryer. *Journal of Food Engineering, 54*(1), 45–58. https://doi.org/10.1016/s0260-8774(01)00184-4.

Gharsallaoui, A., Roudaut, G., Chambin, O., Voilley, A., & Saurel, R. (2007). Applications of spray-drying in microencapsulation of food ingredients: An overview. *Food Research International, 40*(9), 1107–1121. https://doi.org/10.1016/j.foodres.2007.07.004.

Hernandez, J. A., Pavon, G., & Garcıa, M. A. (2000). Analytical solution of mass transfer equation considering shrinkage for modeling food-drying kinetics. *Journal of Food Engineering, 45*(1), 1–10.

Hickey, A. J., & Giovagnoli, S. (2018). Quality by design for particulate systems. *AAPS Introductions in the Pharmaceutical Sciences* (pp. 91–98). https://doi.org/10.1007/978-3-319-91220-2_11

Ishwarya, S. P., & Anandharamakrishnan, C. (2015). Spray-freeze-drying approach for soluble coffee processing and its effect on quality characteristics. *Journal of Food Engineering, 149*, 171–180.

Jafari, S., Salmanzadeh, M., Rahnama, M., & Ahmadi, G. (2010). Investigation of particle dispersion and deposition in a channel with a square cylinder obstruction using the lattice Boltzmann method. *Journal of Aerosol Science, 41*(2), 198–206. https://doi.org/10.1016/j.jaerosci.2009.10.005.

Jafari, S. M., Ghalegi Ghalenoei, M., & Dehnad, D. (2017). Influence of spray drying on water solubility index, apparent density, and anthocyanin content of pomegranate juice powder. *Powder Technology, 311*, 59–65. https://doi.org/10.1016/j.powtec.2017.01.070.

Jafari, S. M., Assadpoor, E., He, Y., & Bhandari, B. (2008). Encapsulation efficiency of food flavours and oils during spray drying. *Drying Technology, 26*(7), 816–835. https://doi.org/10.1080/07373930802135972.

Janiszewska, E., & Witrowa-Rajchert, D. (2009). The influence of powder morphology on the effect of rosemary aroma microencapsulation during spray drying. *International Journal of Food Science and Technology, 44*(12), 2438–2444. https://doi.org/10.1111/j.1365-2621.2009.02025.x.

Karasu, E. N., & Ermis, E. (2019). Determination of the effect of exopolysaccharide (EPS) from Lactobacillus brevis E25 on adhesion of food powders on the surfaces, using the centrifuge technique. *Journal of Food Engineering, 242*, 106–114. Retrieved from https://linkinghub.elsevier.com/retrieve/pii/S0260877418303637.

Khalilian Movahhed, M., & Mohebbi, M. (2016). Spray drying and process optimization of carrot–celery juice. *Journal of Food Processing & Preservation, 40*(2), 212–225.

Knight, P. C. (2001). Structuring agglomerated products for improved performance. *Powder Technology, 119*(1), 14–25. https://doi.org/10.1016/s0032-5910(01)00400-4.

Koç, M., Koç, B., Yilmazer, M. S., Ertekin, F. K., Susyal, G., & Bagdatlioglu, N. (2011). Physicochemical characterization of whole egg powder microencapsulated by spray drying. *Drying Technology, 29*(7), 780–788. https://doi.org/10.1080/07373937.2010.538820.

Koç, M., Güngör, Ö., Zungur, A., Yalçın, B., Selek, İ., Ertekin, F. K., & Ötles, S. (2015). Microencapsulation of extra virgin olive oil by spray drying: Effect of wall materials composition, process conditions, and emulsification method. *Food and Bioprocess Technology, 8*(2), 301–318.

Kostoglou, M., & Karapantsios, T. D. (2003). On the thermal inertia of the wall of a drum dryer under a cyclic steady state operation. *Journal of Food Engineering, 60*(4), 453–462. https://doi.org/10.1016/s0260-8774(03)00076-1.

Largo Ávila, E., Cortés Rodríguez, M., & Ciro Velásquez, H. J. (2015). Influence of maltodextrin and spray drying process conditions on sugarcane juice powder quality. *Revista Facultad Nacional de Agronomía Medellín, 68*(1), 7509–7520. https://doi.org/10.15446/rfnam.v68n1.47839.

Lewicki, P. P. (2006). Design of hot air drying for better foods. *Trends in Food Science and Technology, 17*(4), 153–163. https://doi.org/10.1016/j.tifs.2005.10.012.

Liu, W., Chen, X. D., Cheng, Z., & Selomulya, C. (2016). On enhancing the solubility of curcumin by microencapsulation in whey protein isolate via spray drying. *Journal of Food Engineering, 169*, 189–195. https://doi.org/10.1016/j.jfoodeng.2015.08.034.

Mani, S., Tabil, L. G., & Sokhansanj, S. (2006). Effects of compressive force, particle size and moisture content on mechanical properties of biomass pellets from grasses. *Biomass and Bioenergy, 30*(7), 648–654. https://doi.org/10.1016/j.biombioe.2005.01.004.

McCabe, W. L., Smith, J. C., & Harriott, P. (2005). *Unit operations of chemical engineering* (7th ed.). New York: McGraw-Hill.

Mikli, V., Kaerdi, H., Kulu, P., & Besterci, M. (2001). Characterization of powder particle morphology/Pulbriosakeste morfoloogia kirjeldamine. In *Proceedings of the Estonian Academy of Sciences: Engineering* (pp. 22–35). Tallinn: Academy Publishers.

Moghaddam, A. D., Pero, M., & Askari, G. R. (2017). Optimizing spray drying conditions of sour cherry juice based on physicochemical properties, using response surface methodology (RSM). *Journal of Food Science and Technology, 54*(1), 174–184.

Molina-Sabio, M., & Rodríguez-Reinoso, F. (2004). Role of chemical activation in the development of carbon porosity. *Colloids and Surfaces A: Physicochemical and Engineering Aspect, 241*(1–3), 15–25. https://doi.org/10.1016/j.colsurfa.2004.04.007.

Murrieta-Pazos, I., Gaiani, C., Galet, L., & Scher, J. (2012). Composition gradient from surface to core in dairy powders: Agglomeration effect. *Food Hydrocolloids, 26*(1), 149–158. https://doi.org/10.1016/j.foodhyd.2011.05.003.

Opáth, R. (2014). Technical exploitation parameters of grinding rolls work in flour mill. *Research in Agricultural Engineering, 60*(Special Issue), S92–S97. https://doi.org/10.17221/41/2013-rae.

Patel, R. P., Baria, A. H., & Patel, N. A. (2014). An overview of size reduction technologies in the field of pharmaceutical manufacturing. *Asian Journal of Pharmaceutics, 2*(4), 216.

Paudel, A., Worku, Z. A., Meeus, J., Guns, S., & Van den Mooter, G. (2013). Manufacturing of solid dispersions of poorly water soluble drugs by spray drying: Formulation and process considerations. *International Journal of Pharmaceutics, 453*(1), 253–284. https://doi.org/10.1016/j.ijpharm.2012.07.015.

Pietsch, W. B. (2008). *Agglomeration processes: Phenomena, technologies, equipment*. Weinheim: Wiley.

Pinto, M. R., Paula, D. de A., Alves, A. I., Rodrigues, M. Z., Vieira, É. N. R., Fontes, E. A. F., & Ramos, A. M. (2018). Encapsulation of carotenoid extracts from pequi (Caryocar brasiliense Camb) by emulsification (O/W) and foam-mat drying. *Powder Technology, 339*, 939–946. https://doi.org/10.1016/j.powtec.2018.08.076.

Pua, C. K., Hamid, N. S. A., Tan, C. P., Mirhosseini, H., Rahman, R. B. A., & Rusul, G. (2010). Optimization of drum drying processing parameters for production of jackfruit (Artocarpus heterophyllus) powder using response surface methodology. *LWT - Food Science and Technology, 43*(2), 343–349. https://doi.org/10.1016/j.lwt.2009.08.011.

Ré, M. I. (1998). Microencapsulation by spray drying. *Drying Technology, 16*(6), 1195–1236. https://doi.org/10.1080/07373939808917460.

Reineccius, G. A. (1989). Flavor encapsulation. *Food Reviews International, 5*(2), 147–176. https://doi.org/10.1080/87559128909540848.

Rulkens, W. H., & Thijssen, H. A. C. (2007). The retention of organic volatiles in spray-drying aqueous carbohydrate solutions. *International Journal of Food Science and Technology, 7*(1), 95–105. https://doi.org/10.1111/j.1365-2621.1972.tb01644.x.

Saad, M. M., Barkouti, A., Rondet, E., Ruiz, T., & Cuq, B. (2011). Study of agglomeration mechanisms of food powders: Application to durum wheat semolina. *Powder Technology, 208*(2), 399–408. https://doi.org/10.1016/j.powtec.2010.08.035.

Samaram, S., Mirhosseini, H., Tan, C. P., & Ghazali, H. M.(2014). Ultrasound-assisted extraction and solvent extraction of papaya seed oil: Crystallization and thermal behavior, saturation degree, color and oxidative stability. *Industrial Crops and Products, 52*, 702–708. https://doi.org/10.1016/j.indcrop.2013.11.047.

Schubert, H. (1987). Food particle technology. Part I: Properties of particles and particulate food systems. *Journal of Food Engineering, 6*(1), 1–32. https://doi.org/10.1016/0260-8774(87)90019-7.

Shenoy, P., Viau, M., Tammel, K., Innings, F., Fitzpatrick, J., & Ahrné, L. (2015). Effect of powder densities, particle size and shape on mixture quality of binary food powder mixtures. *Powder Technology, 272*, 165–172. https://doi.org/10.1016/j.powtec.2014.11.023.

Sharma, M., Kadam, D. M., Chadha, S., Wilson, R. A., & Gupta, R. K. (2013). Influence of particle size on physical and sensory attributes of mango pulp powder. *International Agrophysics, 27*(3), 323–328.

Soottitantawat, A., Bigeard, F., Yoshii, H., Furuta, T., Ohkawara, M., & Linko, P. (2005). Influence of emulsion and powder size on the stability of encapsulated d-limonene by spray drying. *Innovative Food Science & Emerging Technologies, 6*(1), 107–114. https://doi.org/10.1016/j.ifset.2004.09.003.

Stange, U., Scherf-Clavel, M., & Gieseler, H. (2013). Application of gas pycnometry for the density measurement of freeze-dried products. *Journal of Pharmaceutical Sciences, 102*(11), 4087–4099. https://doi.org/10.1002/jps.23723.

Szulc, K., & Lenart, A. (2013). Surface modification of dairy powders: Effects of fluid-bed agglomeration and coating. *International Dairy Journal, 33*(1), 55–61. https://doi.org/10.1016/j.idairyj.2013.05.021.

Turchiuli, C., Jimenez Munguia, M. T., Hernandez Sanchez, M., Cortes Ferre, H., & Dumoulin, E. (2014). Use of different supports for oil encapsulation in powder by spray drying. *Powder Technology, 255*, 103–108. https://doi.org/10.1016/j.powtec.2013.08.026.

Valous, N. A., Gavrielidou, M. A., Karapantsios, T. D., & Kostoglou, M. (2002). Performance of a double drum dryer for producing pregelatinized maize starches. *Journal of Food Engineering, 51*(3), 171–183. https://doi.org/10.1016/s0260-8774(01)00041-3.

Vogel, C., Scherf, K. A., & Koehler, P. (2018). Effects of thermal and mechanical treatments on the physicochemical properties of wheat flour. *European Food Research and Technology, 244*(8), 1367–1379. https://doi.org/10.1007/s00217-018-3050-3.

Walton, D. E. (2000). The morphology of spray-dried particles a qualitative view. *Drying Technology, 18*(9), 1943–1986. https://doi.org/10.1080/07373930008917822.

Walton, D. E., & Mumford, C. J. (1999). The morphology of spray-dried particles: The effect of process variables upon the morphology of spray-dried particles. *Chemical Engineering Research and Design, 77*(5), 442–460.

Wang, S., & Langrish, T. (2009). A review of process simulations and the use of additives in spray drying. *Food Research International, 42*(1), 13–25. https://doi.org/10.1016/j.foodres.2008.09.006.

Yohannes, B., Liu, X., Yacobian, G., & Cuitiño, A. M. (2018). Particle size induced heterogeneity in compacted powders: Effect of large particles. *Advanced Powder Technology, 29*(12), 2978–2986. https://doi.org/10.1016/j.apt.2018.09.020.

Chapter 3
Adhesion of Food Powders

Ertan Ermiş

3.1 Introduction

In food industry, particle adhesion and cohesion has particular importance in many applications such as coating with powders, dosing of powders, surface cleaning of the process equipment, discharge from the storage units, mixing/blending, packaging and new product development. Therefore, control of adhesion and cohesion forces in food powders is essential to design the processes and to improve the process efficiency.

Adhesion is the state in which a powder particle held on a solid surface by different interfacial forces (i.e. van der Waals, electrostatic and capillary forces) and mechanical interlocking action, while cohesion describes the affinity of one particle to another. Adhesion can be managed by controlling numerous factors such as particle size, shape, surface morphology, moisture content and chemical composition.

While the adhesion of particulates or soft solids on surfaces is desired in some food applications, adhesion of unwanted material and their accumulation to form fouling deposits is not wanted in some processing stages. This unwanted adhesion may reduce process efficiency and lower product quality as well as compromising hygiene and may result in cross-contamination.

Different techniques have been developed to apply powder coating onto the surfaces. Since the adhesion of particles on the surfaces is not well understood, various methods have been studied to determine the adhesion or removal behaviour of food powders on surfaces. In this chapter, the adhesion of food powders, the recent developments in determination of particle adhesion and evaluation of the performance of the test methods developed are given.

E. Ermiş (✉)
Food Engineering Department, Faculty of Engineering and Natural Sciences, Istanbul Sabahattin Zaim University, Istanbul, Turkey

E. Ermiş (ed.), *Food Powders Properties and Characterization*, Food Engineering Series, https://doi.org/10.1007/978-3-030-48908-3_3

3.2 Adhesion of Food Powders

The adhesion can be defined as the attraction between two solid material having a common contact surface which is produced by intermolecular attractive forces within short distances (Petean and Aguiar 2015). Adhesion plays an important role in many applications such as powder coating and processing of cohesive powders. Powder coating refers to the adhesion of food powder onto the surface of food products (Ermis et al. 2011). Seasoning powders are often added to improve the appearance and taste of the food products. Adhesion of seasonings to snack foods should be consistent and to obtain uniform and even distribution of seasonings on the surfaces, most manufacturers apply excess coating powder. This may lead to fugitive build-up on the surfaces of processing equipment and this may result in giving break in production for cleaning of the processing equipment (Adhikari et al. 2001; Ermis et al. 2011). The powder might become lumpy over the equilibrium moisture and temperature especially for milk and fruit powders due to the existence of low molecular weight sugars in high concentration (i.e. glucose, fructose, and sucrose) (Adhikari et al. 2001).

For most of the snack foods, seasonings are applied together with oil or immediately after frying process. It has been reported that increasing surface oil content resulted in increase of adhesion (Enggalhardjo and Narsimhan 2005). Studying adhesion mechanisms and measuring of the adhesive force is needed to better understand the powder behavior to diminish powder waste and to increase efficiency in powder coating processes (Zafar et al. 2014).

Numerous factors such as flowability, relative humidity, particle size and particle shape can influence the adhesion of powders to the surfaces. The flowability and dispersibility of fine powders usually become poor when particle-particle and particle-surface adhesion strength is increasing (Zafar et al. 2014). Powder flowability is affected by particle size (Teunou et al. 1999). In general small particles having high surface to volume ratio are more cohesive with poor flowability. Free-flowing powders are preferred for uniform (evenly dispersed) powder coating (Khan et al. 2012). Powder with free-flowing property increase the transfer efficiency and dispersion over the food substrate surface. However, cohesive powders may produce better adhesion due to higher percent side coverage on the target surface (Sumawi and Barringer 2005).

It has been reported that particle size has remarkable effect on adhesion strength and wrap-around effect (percent side coverage) during coating because of stronger van der Waals forces per unit of mass (Halim and Barringer 2007; Buck and Barringer 2007; Ermis et al. 2011). As the particle size increases, the percent side coverage decreases and hence weaker adhesion occurs between particle and surface due to lower charge-to-mass ratio (Halim and Barringer 2007). Attention has been paid to adhesion properties of fine particles in process engineering due to the fact that those properties may cause unwanted effects such as blockage, detachment or poor flowability leading to reduced efficiency in production processes (Adhikari et al. 2001). Adhesion of fine particles to surfaces is closely relateds to industrial hygiene and air pollution (Petean and Aguiar 2015).

Particle shape and the surface area of particles are other factors affecting adhesion to surfaces (Karasu and Ermis 2019; Buck and Barringer 2007; Nussinovitch 2017). It has been reported that flake shaped particles adhere better than cubic and dendritic shaped ones, particularly for bigger particles than 200 μm (Miller and Barringer 2002; Niman 2000). In addition, chemical composition, moisture content and water activity of both powder material and substrate are affecting adhesion. One of the most important factors affecting particle adhesion is surface oil/liquid content. The effectiveness of coating increased with increasing oil content (Miller and Barringer 2002). The type of oil and it's composition affects adhesion behaviour of food powders due to the differences in viscosities (Enggalhardjo and Narsimhan 2005). The surface topology, porosity and roughness are also other factors affecting adhesion of particulates. If the particle size is smaller than the pore size of the surface, mechanical interlocking may take place (Bowling 1988). Substrate temperature is reported as another factor affecting adhesion of food powders in the presence of surface oil. Decrease in surface temperature yields lower adhesion (Buck and Barringer 2007).

Adhesion of food powders in food containers and packages may affects consumers' acceptability adversely. Adhesion of food powders in packaging causes deformation of the product surface. Unwanted powder adhesion in packages may cause consumer antipathy to the product (Kilcast and Roberts 1998; Adhikari et al. 2001).

Various methods have been used in coating food substrates (i.e. snack food) with powder materials such as seasoning powders. Tumble drum and conveyor belt techniques are the most common methods for seasoning coating. Another emerging technique, electrostatic powder coating is one of those methods which can be applied in industrial applications (Halim and Barringer 2007). In this method, powder particles form a powder cloud and disperse evenly on the food surface by applying electrostatic forces which result in attraction between charged particles and food surfaces due to the high charge-to-mass ratio of the small particles (Halim and Barringer 2015). Electrostatic powder coating was reported as a promising technique for smaller size fractions (particularly below 100 μm) when there is no oil or liquid on the surface of the substrate (Buck and Barringer 2007).

To better understand the behaviour of food powders, the adhesion forces and factors affecting adhesion need to be studied in detail. Several experimental, theoretical and numerical studies have been carried out to measure and to better understand the particle adhesion (Zafar et al. 2014).

3.3 Powder Coating Systems

Coating with food powder involves the application of particulates on the surface of food product to achieve new functionalities such as improving sensory attributes (i.e. color and flavour) for increased consumer acceptability; improving nutritional quality (i.e. application of vitamins, mineral sor bioactives); improving flowability by applying anticaking agents (e.g. powdered cellulose on graded cheese) or to coat with antimicrobial agents for enhancing shelf-life (Dhanalakshmi et al. 2011; Yousuf and Barringer 2007; Khan et al. 2012; Elayedath and Barringer 2002). Therefore, effective targeting of coating material and even distribution of powder particles on the surface of the food product is desired in food industry (Khan et al. 2012). Powder coating should be done properly to spread the coating material across the target evenly with high transfer efficiency. The adhesion of powder material is very important to keep the particles on the food surfaces. To fixate powder particles to the product surface, oil or hydrocolloid solutions first sprayed on the target surface to promote adhesion of particles.

3.3.1 Coating in the Gravity Drum (Tumble Drum Coating System)

Tumble drum coating is a common and widely used method to apply coating if the food product can be tumbled (Biehl and Barringer 2003). Tumble drums are horizontal cylinders with a degree of inclination to allow product flows from one end to the other (Hui 2006). They are designed in a way to create folding-in action which cascades down the surface of the food products. The seasoning particles are carried in a direction opposite the cascading layer (Denis et al. 2003). Uniform distribution of the flavouring powder is achieved by exposing the entire base product to the flavour spray or powder application within the coating zone while mixing the food product as a result of tumbling action (Elayedath and Barringer 2002). Powder density and feed rate of powder and the volume of the food product in the tumble mixer and the rotational speed affects coating efficiency (Wong et al. 2005). In most of the cases, a liquid media such as oil is sprayed onto the product in the tumbler to help the food powder adhere to the surfaces (Ermis 2011).

3.3.2 Belt System

The belt coating systems are in general facilitate single-side coating of products like salted crisps, crackers or pretzels (Dreier 1991). Different types of conveyor can be used such as vibratory, open wire belt, closed fabric belt, or other selected types. In this method, a given amount of powder is applied on the conveyor belt entering the coating zone. In this method, products need to be turned around when coated on a conveyer belt to evenly distribute the powder material across the target surface (Lusas and Lloyd 2001; Riaz 2015). However, coating all sides of a three-dimensional product uniformly is challenging when this technique is used. The coating efficiency, which can be defined as the amount of powder that adheres to the target compared to the total mass of powder, is relatively low in this method. Therefore, more powder material is used than the required amount to achieve desired coating efficiency (Khan et al. 2012).

3.3.3 Electrostatic Coating

Electrostatic powder coating technique uses electrostatic charges to improve the adhesion strength of powder particles onto the food substrates. A high voltage electrode attached corona gun is used to apply an electrical charge to the powder material dispersed or blowed over the surface of the substrate to be coated (Fig. 3.1). High voltage electrode generates an electric field and produce negative charges towards the particles. The grounded substrate attracts the charged particles (Khan et al. 2012).

It has been reported that this technique improves efficiency of particle coating. Since the applicator is sealed, there is no powder loss to the environment which result in reducing wastage. It has been reported that 20% less oil is needed for adhesion when using electrostatics and increasing relative humidity reduces electrostatic adhesion (Khan et al. 2012).

3.3.4 Atmospheric Plasma Coating System

Atmospheric pressure plasma discharge has been used in powder coating to the surfaces in different industries such as ceramic and automotive. The radicals, ions and active molecules are produced from a gas feed by applying high voltage of electricity. Those active molecules, ions and free radicals activate the surface of the

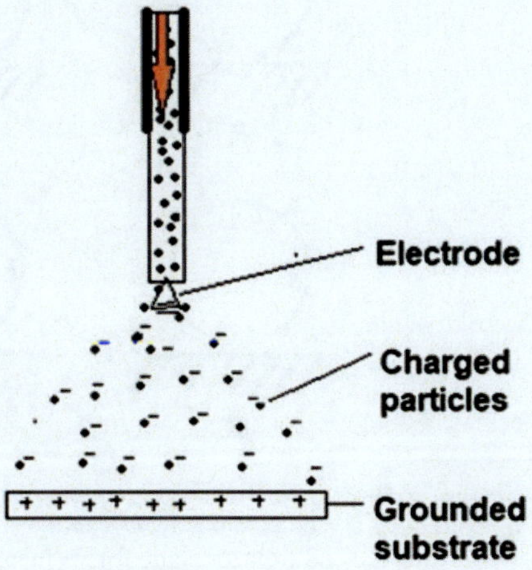

Fig. 3.1 Schematic view of electrostatic powder coating [Adapted from Prasad et al. 2016]

substrate and the particle to enhance the adhesion of powder material to the surface of the substrate. Even though there have been some researchers studying the effect of plasma discharge on food powder coating (Suganya et al. 2018), there is still a lack of information in this field.

3.4 Adhesion Mechanisms

The adhesion mechanisms and the type of adhesion forces between the particle-surface vary depending on the chemical and physical properties of solid material building the surface and particles and existence of any liquid layer between the solid surfaces. The liquid layer (i.e. oil or hydrocolloid solution) improves the interaction between liquid and soluable solids which promotes adhesion. The chemical composition and molecular structure plays an important role in adhesion behaviour. Depending on the chemical composition and molecular structure, food materials show different properties such as hygroscopic and hydrophobic properties. Based on these behaviours, food materials can be divided into three groups. These are amorphous, crystalline and semi-crystalline materials. Most of the food powders show hydroscopic (water soluable) behaviour. Some foods which are rich in fats and oils may behave as hydrophobic. (Dopfer et al. 2013). Particle shape, particle size and contact geometry affect the magnitude of adhesion force measured. For fine particles (roughly 1–10 micron), attractive surface forces dominates over inertial forces (increase in surface charge to mass ratio) and this increases adhesion of particles (Wanka et al. 2013).

Different adhesion mechanisms take place between the particle and the surface of food materials depending on the chemical composition and structure (Fig. 3.2).

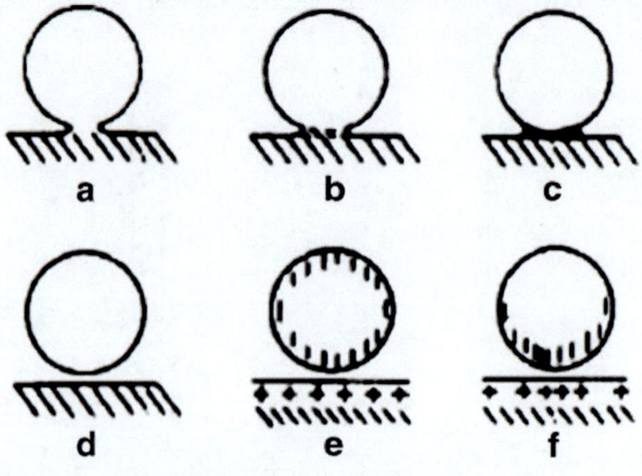

Fig. 3.2 Adhesion mechanisms between a particle and a surface. a-chemical bonding (sintering), b-bonding by crystallised solute, c- capillary forces or liquid bridge (viscous bonding), d-van der Waals forces, e and f-electrostatic forces [adapted from Schubert 1987]

The total adhesion force is assumed to be the sum of van der Waals, F_{vdw} (significant for fine powders), capillary, F_{cap} (dominant when there is a liquid layer) and electrostatic, F_{el}, (dominant when the particles are highly charged) forces among others (Ermis et al. 2011; Zafar et al. 2014; Salazar-Banda et al. 2007). Van der Waals forces increase as particle size decreases. The sum of these forces can be assumed as total adhesion force (Salazar-Banda et al. 2007):

$$F_{ad} = F_{vdW} + F_{cap} + F_{el} \tag{3.1}$$

The adhesion force between particles and between particles and surfaces is usually due to van der Waals interactions if the powder and surface are dry and without any chemical bonds and electric field (Salazar-Banda et al. 2007). It has been reported that the electrostatic forces are much smaller than the force of van der Waals if there is an compression force acting on the particle (Petean and Aguiar 2015).

Particle size has a significant influence on the cohesive and adhesive strength of a particulate system. Due to high surface to volume ratio of micro and nano-particles, they are highly susceptible to electrostatic or molecular interactions while van der Waals forces become significant below 10 micron size (Rumpf and Knepper 1962). Powders having the median particle size over 200 micron are considered as free-flowing and finer particles tend to stick together and show poor flowability due to cohesion (Teunou et al. 1999). For particles below 1 micron size, deformation may occur due to molecular forces and this leads to increase contact area which result in more intimate contact (Cyprien and Ludwik 1999). Adhesion mechanisms of particles have been studied previously and reviews have been reported in detail (Bowling 1988; Kendall and Stainton 2001; Kumar et al. 2013; Schubert 1987; Adhikari et al. 2001).

3.4.1 Electrostatic Forces

When electrically nonconducting particles come in contact, they are attributable to different contact potential values. When particles possess excess opposing charges, electrostatic adhesion may take place (Schubert 1987). The electrostatic adhesion forces for a nonconducting particle, F_{el} can be calculated using Coulomb's law as:

$$F_{el} = \frac{\pi q_1 q_2 d_1^2 d_2^2}{\varepsilon_r \varepsilon \left(d_1 + d_2 + 2x\right)} \tag{3.2}$$

where q_1 and q_2 are the electric charges of the spherical particles (Coulomb.m^{-2}), εr is relative dielectric constant (dimensionless) and ε is absolute dielectric constant (Coulomb2.N1.m^{-2}) of the surrounding medium, x (m) is the distance of separation between the particles. d_1 and d_2 are diameters of the spheres (Schubert 1987).

For electrically conducting particles, the adhesion force, F_{el} attributable to contact potential between two rigid spheres is given by:

$$F_{el} = \frac{\pi \varepsilon_r \varepsilon U^2 d_1 d_2}{2(d_1 + d_2)x} \tag{3.3}$$

where, F_{el} is force of cohesion/adhesion between the two spherical particles or between a sphere and a plane, U is contact potential (N.m.s^{-1} or volt) (varies between 0.1 and 0.7 V) (Schubert 1987).

3.4.2 Van der Waals Forces

Van der Waals forces (F_{vdW}) are generated by molecules positioned in close contact distances. These forces are kind of electrostatic forces generated as a result of temporary load shifts of the molecules due to plastic or visco-elastic deformation of particles. Lifshitz (1955) and Hamaker (1937) linked these forces to the contact zone characteristics between two surfaces of circular plates (see Eq. (x)) (Dopfer et al. 2013).

Lifhitz:

$$F_{vdW} = \frac{h\overline{w}}{32\pi h_s^3} x^2 \tag{3.4}$$

Hamaker (for $h_s < 150$ nm):

$$F_{vdW} = \frac{H}{24h_s^3} x^2 \tag{3.5}$$

x is the diameter of the circular contact area of the two particles, h_s is the separation distance between the two particles or parallel surfaces. $h \cdot \varpi$ is the Lifshitz–Van der Waals constant (10^{-20}– 10^{-18} J) and H is the Hamaker constant which depends on the materials in contact and surrounding liquid (10^{-19} – 10^{-18} J).

The distance between the particles and the contact area strongly affect the vdW forces. An increase in contact area and/or a decrease in the distance between their surfaces increases vdW forces. The changes in these geometrical properties are linked to the degree of plastic and visco-elastic deformation of the particles which are depending on mechanical particle properties (i.e. viscosity and elasticity) (Dopfer et al. 2013). Physico-chemical properties of amorphous water-soluble powder materials (i.e. maltodextrin) are affected by the concentration of water. Migration of water into the structure of an amorphous particle may lead to decrease in glass transition temperature and change in some of the rheological and mechanical properties such as viscosity and elasticity (Dopfer et al. 2013). This behaviour may result in visco-elastic deformation of particulates and hence improved adhesion by increasing contact area between the particles and hence decreasing distance between the particle surfaces which leads to an increase in Van der Waals forces (Dopfer et al. 2013).

The F_{vdw} can be calculated for two spheres (ideally smooth and rigid spheres) of diameter d_1 (*m*) and d_2 (*m*) separated from each other by distance x (*m*) (Adhikari et al. 2001), as:

$$F_{vdW} = \frac{E_p d_1 d_2}{16\pi \left(d_1 d_2\right) x^2} \tag{3.6}$$

Ep, the Van der Waals' interaction energy ($10^{-19} - 10^{-18}$ J). In the case of particles are in close contact with each other, Van der Waals' force (N) is becoming stronger. Eq. (3.6) can be applied to both particle-particle and particle-plane interfaces (Schubert 1987).

3.4.3 *Capillary Forces and Liquid Bridges*

This kind of adhesion forces are generated by low viscosity liquid bridges between particle-surface systems. It has been reported that capillary forces are becoming dominant over electrostatic and vdW forces when there is a low viscosity liquid layer in between a particle (Fig. 3.3) and a surface of a substrate (Dopfer et al. 2013; Wang et al. 2017). The capillary forces between two particles could be calculated using different approaches. Rabinovich et al. (2005) used a total energy approach to determine capillary forces. Van der Waals forces exhibit a stronger decrease when compared to capillary forces in the case of increasing separation distance between particulate systems. A number of publications cover adhesive forces generated between particles having liquid bridges in between (Dopfer et al. 2013; Payam and Fathipour 2011; Megias-Alguacil and Gauckler 2009; Simons 2007).

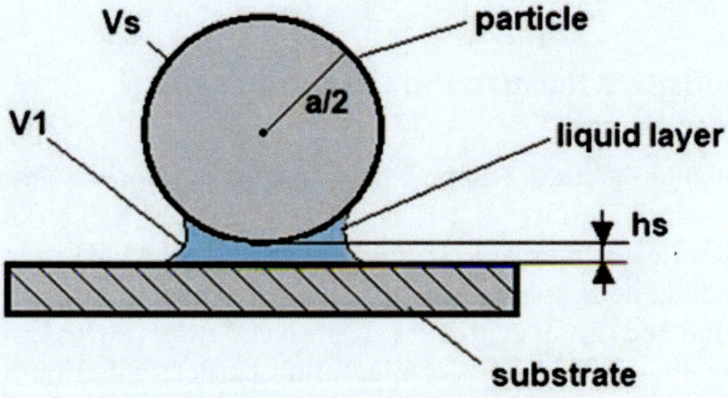

Fig. 3.3 Adhesion of a particle due to capillary forces. (V1/Vs is liquid to solid volume ratio, a is particle diameter, hs is separation distance)

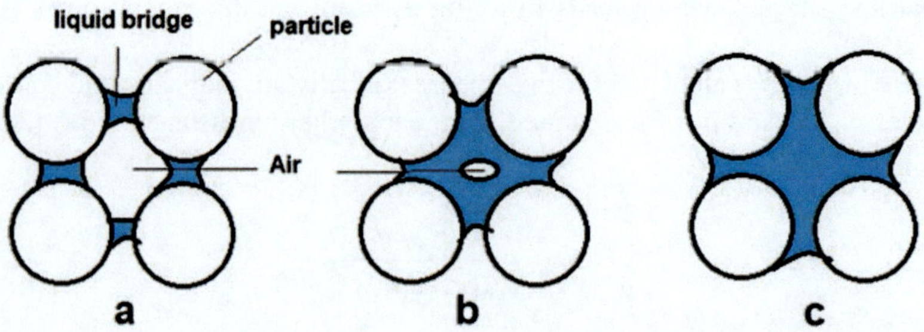

Fig. 3.4 Schematic illustration of immobile liquid bridges. a-pendular state, b-funicular state, c-capillary state [adapted from Peleg (1977)]

Liquid bridges between particle-surface or particle-particle improve adhesion and cohesion phenomena. These liquid bridges can be classified as mobile and immobile and immobile ones can be divided into three groups: pendular, funicular and capillary state (Fig. 3.4) (Adhikari et al. 2001).

Liquid is absorbed at the surface of the granules as well as it penetrate into the supra-molecular and capillary structure of the particles. This liquid transfer by the time affects the adhesion caused by capillary forces. In general, when the relative humidity of the storage unit where powder material kept is above 70%, capillary forces predominate (Halim and Barringer 2007).

3.4.4 Mechanical Inter-Locking

Interlocking or folding action may occur between fibrous, bulky, and flaky particles having irregular surface characteristics and this may result in "form-closed" bonds (Pietsch 2003). This mechanical bonding (mechanical interlocking/matting/meshing between the particles) occurs especially when the temperature of the material rise to certain degrees to cause molecular flow into each other (Griffith 1991).

3.5 Methods to Measure Particle Adhesion

Various test methods and test equipments have been developed to measure adhesive forces between the interface of paticle-particle and particle-surface (Table 3.1). Some examples are given as Atomic Force Microscopy (AFM), centrifugal detachment, electric field detachment, aerodynamic detachment, drop test, impact tester and vibration method (Zafar et al. 2014; Karasu and Ermis 2019; Ermis et al. 2011). Most of the methods developed for measuring particle adhesion are either time-consuming or experimentally demanding or not practically applicable. Since most

Table 3.1 Adhesion measurements methods [adapted from Ermis et al. 2011]

Method	Operation	Advantages	Disadvantages
Centrifugation	• Tangential detachment • Image analysis/ weighing before and after testing	• Accurate and repeatable • Simple and well established • Good statistics	• Similar size of particles need to be used • Cohesive powders might be problematic • Time consuming
Aerodynamic detachment	• Detachment at angled positions by air jet • Image analysis/ weighing before and after testing	• Flexible on substrate size • Good statistics	• Particle-particle collisions and particle layers on the surface need to be considered • Velocity of air is high
Hydrodynamic detachment	• Detachment caused by liquid stream • Image analysis/ weighing before and after testing	• Flexible on substrate shape • Good statistics	• For only insoluable particles
Impact separation	• Detachment by impact on particles surface with adhered on opposite side • Image analysis/ weighing before and after testing	• Test method is not complicated, easy to conduct • Relatively low capital cost • Good statistics	• Fine particles need high g forces • Existence of a liquid layer cause capillary force and hence more impact force required • May damage surface-particle systems at high impact forces
Vibration	• Detachment by acoustic transducer • Image analysis/ weighing before and after testing	• May be used with liquids • Good statistics	• May damage surface-particle systems at high vibration forces • Plastic deformation may occur
Electrostatic detachment	• Voltage is applied between electrodes • Image analysis/ weighing before and after testing	• Relatively fast method • Good statistics	• For only charged particles
Atomic force microscope (AFM)	• Optical beam deflection is proportional to deflection of cantilever • Piezoelectric sensor measures the force needed to detach the particle	• Precision and control is high • Same particle can be tested at different conditions • Different tips distance • Short contact time • Measures attractive as well as removal force	• Long time needed for sample preparation • Cohesive powders may cause difficulties • Capital cost is high • Enough number of tests need to be conducted for statistics • Poor statistics

of the techniques are not usable for routine applications, researchers have been focusing on a simple alternative method.

The techniques studied so far had different mechanisms and approaches to measure particle adhesion. Therefore they produce varios measure of adhesion. Some examples to the approaches are surface contact measurement and particle deposition and detachment (Hu et al. 2010). Surface topography, particle physical properties (shape, size, surface structure, particle density etc.) and contact geometry affect the measured adhesion force (Packham 2003). Temperature and moisture are extrinsic parameters affecting the magnitude of adhesion force (LaMarche et al. 2010).

Some of the methods such as Atomic Force Microscopy (AFM) may be costly and focus on single particle only. However some applications involve bulk powder treatment and a need for rapid and reliable test method was arised. For this reason the drop test method (Zafar et al. 2014) and impact adhesion tester (Ermis et al. 2011) have been developed and studied to be able to determine adhesion of powders attached to the surfaces in bulk to mimic the industrial applications such as flavor coating. In these test methods particles which adhered to the surface of a substrate are attached on a sample holder and they are subjected to a tensile force/impact force by impacting the stub against a stopper by dropping it from a set height. Vibration method was studied by (Vahdat et al. 2013) and Hopkinson bar technique was applied by (Wanka et al. 2013). A review of detachment methods was presented by (Petean and Aguiar 2015).

3.5.1 Colloidal Probe Technique/Atomic Force Microscopy (AFM)

The AFM technique have been used by different researchers to characterize particle adhesion (Kappl and Butt 2002; Beach et al. 2002; Duri et al. 2013). However, it only measures single particle adhesion and too many tests need to be conducted due to irregularities of the particles to have reliable data, a limiting factor for representative number of particles that can be investigated (Wanka et al. 2013). Each single particle has to be attached manually to the end of a microcantilever making it not suitable for measuring bulk powder adhesion. In addition, the equipment is expensive and the measurement is time-consuming and for obtaining reliable data, several measurements need to be conducted for irregular particles because of variations of contact geometries and local surface properties.

3.5.2 Centrifuge Method

The adhesion between particle and surface of a substrate can be determined by measuring the force generated by angular speed of rotation of a particle adhered to the surface of a substrate placed in a rotor of a centrifuge with a defined particle mass at a known distance to the rotation centre (Ermis et al. 2011; Salazar-Banda et al. 2007; Karasu and Ermis 2019; Knoll et al. 2015). Varying rotational speeds can be applied on the substrate where particles are attached to determine particle detachment from the surface. In this method, the necessary force to detach the particles varies proportionally to particle size, particle shape and surface characteristics. The adhesion force, generated in the detachment direction, acting on a single particle can be calculated by Eq. 3.7 (Salazar-Banda et al. 2007).

$$F_{cen} = m\omega d^2 dc \tag{3.7}$$

where m is the mass of the particle, wd is the angular speed necessary for detachment and dc is the distance between the sample and the rotor centre (rotational axis).

By using centrifuge method, adhesion forces of a large number of particles can be measured simultaneously even though the time needed for each test might be long due to spin up/down action for each acceleration.

3.5.3 *Electrostatic/Electric Field Detachment*

Another method studied to characterise particle adhesion is electrostatic detachment (Takeuchi 2006). In this method, electric field is used to remove particles from the surface. This method can be used only for electrically conductive particles which limits its use.

3.5.4 *Aerodynamic/Hydrodynamic Detachment*

The earodynamic method was employed to measure particle adhesion by using a gas or air stream across the surface (Enggalhardjo and Narsimhan 2005; Shukla and Henthorn 2009). It was reported that in this method, particle–particle collisions may occur and the drag force may be influenced by the close proximity of the particles and this may lead to breakage of the particles. This technique uses controlled (or estimated) shear stress conditions (i.e. fluid flow over the surface) and monitor the response of the particle (Burdick et al. 2005). This technique is limited to insoluable particles and surfaces when liquid is used.

3.5.5 *Inertial Detachment*

Wanka et al. (2013) developed a method to determine particle adhesion of fine powders on a surface based on the acceleration generated by the free end of a Hopkinson bar (a long slender titanium bar) which a schock pulse is excited on one end by a bullet from an air gun. In this technique, high acceleration (around 500,000 g) of the bar retract the surface and particle detachment take place due to their inertia. The bar is excited mechanically to provide the acceleration necessary to detach the particles from the surface. This happens when the adhesion force of the particle exceeds by the inertial force provided by the bar (Wanka et al. 2013).

$$F_{ad} = F_{detach} = \rho \frac{4}{3} \pi R^3 a \tag{3.8}$$

where ρ is the particle density, R the particle radius, and a the acceleration of the surface.

3.5.6 *Impact/Drop Detachment*

A bench size impact tester was designed by Ermis et al. (2011) to determine the magnitude of adhesion forces between particles and surface. The tester had a vertically moving platen which was reported to generate upto around 100 g-force depending on the drop height. Similarly, a drop test method was designed and used by Zafar et al. (2014). The principle of this test include particle detachment by the momentum of the particles caused by the sudden deceleration of the substrate attached to an aluminium stub inside a glass tube. For both techniques, the force required to remove a given fraction of adhering particles from the surface of the substrate is calculated by Newton's second law of motion, using a calculated particle mass (based on size, density and acceleration of the particle). Detached particles can be investigated either using a laboratory balance based on weigh loss or by microscopy and image analysis.

For a critical particle size, the detachment take place above the threshold force required to detach the particle and they remain on the surface below the threshold force required to detach the particle. The force needed to detach a particle can be determined using Newton's second law of motion. The variables need to be estimated are particle size, acceleration and particle density. (Zafar et al. 2014) have tested various material such as silanised glass beads, Avicel, α-lactose monohydrate and starch using a drop test apparatus based on microscopic observations of number of particles. The measurement approach is similar to that proposed by (Ermis et al. 2011). They designed and introduced impact adhesion tester (IAT) to be used in food powder adhesion on food surfaces. On impact, a tensile force is generated on the particles, causing detachment action due to the balance between the tensile force and the adhesion strength. The amount of particles detached was calculated by measuring the weight of the coated substrate before and after impact. The principle of both techniques rely on the particle detachment by the momentum of the particles on the substrate while applying varying deceleration. The detachment force can be determined using Newton's second law of motion. Particle mass, based on its volume and density as well as particle deceleration need to be determined to calculate the force needed tp detach a particle.

3.5.7 Vibration Technique

Additionally, it was reported that vibration caused rolling of particles on the surface before detaching and this behaviour reduced the adhesive force between the particles and surface (Kobayakawa et al. 2015).

3.5.8 Resonance Frequency/Vibration

In resonance frequency method, a rocking motion is generated by air acoustic transducer or ultrasonic transducer connected to flat surface (Vahdat et al. 2013). Resonance frequency (excitation by a short acoustic pulse) is used in this method to bring about the rocking motion on the surface where particle has been adhered.

The vibration technique as an alternative method has been developed and studied by (Kobayakawa et al. 2015). A piezoelectric vibrators are employed in this method to generate sinusoidally alternating stress to detach particles from the surface of the substrate. The amplitude of vibration can be controlled and detached particles are analyzed using a high-speed microscope camera. In this method the vibration may flatten the asperities and cause an intensification of the adhesion force (Ripperger and Hein 2008).

Wanka et al. (2013) used Hopkinson bar to determine particle adhesion. The detachment of particles are monitored by an optical microscope while subjecting fine particles to around 500,000 g acceleration. These kind of techniques are limited to test fine particles' adhesion ranging from 3 to 20 μm. One disadvantage of vibrational techniques might be a possible plastic deformation and thus damage to surface and particles at high vibration force.

3.5.9 Mathematical Models

For estimation of the adhesion strength between the particles and the surface, different mathematical models have been studied to describe the contact mechanics of particles. The most widely used methods are the theories of Johnson–Kendall–Robert (JKR) (Johnson et al. 1971) and Derjaguin–Muller–Toporov (DMT) (Derjaguin et al. 1975). The work of adhesion (WA) is defined as the amount of reversible work required to separate a unit area of the surfaces from contact to infinity (Derjaguin et al. 1975).

JKR theory is based on the Hertz analysis, with the addition of adhesion energy modifying the contact area and requiring a pull-off force, which is related to the surface energy of the contact, to detach the adhered particles. This theory consider all short range contact forces exist within the contact area evenly distributed. This approach is more appropriate for soft materials with significant adhesive forces (Johnson et al. 1971).

DMT theory considers molecular attraction between particles acting outside the contact area (estimation of non-contact forces). The DMT model describes weak attraction between stiff materials (Derjaguin et al. 1975).

Both models assume elastically deformable solids in contact with either sphere–sphere or sphere–plane geometries. These models predict a pull-off force necessary to separate the particle from another particle or from a surface.

$$F_{ad} = 2\pi R W_A \left(\text{DMT}\right) \tag{3.9}$$

$$F_{ad} = \frac{3}{2}\pi R W_A \left(\text{JKR}\right) \tag{3.10}$$

where R is the radius of the spherical particle in contact with a plane surface or with another particle. Since work is required to separate the solid surfaces in contact, molecular adhesion can be depicted in terms of energy (Petean and Aguiar 2015).

Based on to the DMT and JKR models, mechanical properties and the geometry of the particle and substrate defines the particle–substrate adhesion. However, the particle– substrate adhesion is also affected by some other factors (i.e. the surface roughnesses of the particle and substrate; the electrostatic charge and the relative humidity of the environment) (Liu et al. 2011). Comprehensive reviews on the fundamentals of contact mechanics are given elsewhere.

3.6 Conclusion

The theoretical and experimental studies have been done about adhesion mechanisms of particulate systems up to date suggested that the material properties such as hydrophobicity, hygroscopicity, mechanical/rheological properties and particle properties (size, shape and surface properties) need to be better understood to describe and better understand complex adhesion and cohesion behaviour of food powders. Most of the research work has been done in the recent years on particle adhesion has been focused on single particle/cell adhesion. However, there is still research work to be done about adhesion of food particles in bulk onto food surfaces at full production scale operations (i.e. snack food manufacturing).

References

Adhikari, B., Howes, T., Bhandari, B., & Truong, V. (2001). Stickiness in foods: A review of mechanisms and test methods. *International Journal of Food Properties, 4*, 1–33. https://doi.org/10.1081/JFP-100002186.

Beach, E. R., Tormoen, G. W., Drelich, J., & Han, R. (2002). Pull-off force measurements between rough surfaces by atomic force microscopy. *Journal of Colloid and Interface Science, 247*, 84–99.

Biehl, H. L., & Barringer, S. A. (2003). Physical properties important to electrostatic and nonelectrostatic powder transfer efficiency in a tumble drum. *Journal of Food Science, 68*, 2512–2515.

Bowling, R. A. (1988). A theoretical review of particle adhesion. *Particles on Surfaces, 1*, 129–142.

Buck, V. E., & Barringer, S. A. (2007). Factors dominating adhesion of NaCl onto potato chips. *Journal of Food Science, 72*(8), E435–E441.

Burdick, G. M., Berman, N. S., & Beaudoin, S. P. (2005). Hydrodynamic particle removal from surfaces. *Thin Solid Films, 488*, 116–123.

Cyprien, G., & Ludwik, L. (1999). On stickiness. *Physics Today, 52*(11), 48–52. https://doi.org/10.1063/1.882884.

Denis, C., Hemati, M., Chulia, D., Lanne, J. Y., Buisson, B., Daste, G., & Elbaz, F. (2003). A model of surface renewal with application to the coating of pharmaceutical tablets in rotary drums. *Powder Technology, 130*, 174–180.

Derjaguin, B. V., Muller, V. M., & Toporov, Y. P. (1975). Effect of contact deformations on the adhesion of particles. *Journal of Colloid and Interface Science, 53*, 314–326.

Dhanalakshmi, K., Ghosal, S., & Bhattacharya, S. (2011). Agglomeration of food powder and applications. *Critical Reviews in Food Science and Nutrition, 51*, 432–441.

Dopfer, D., Palzer, S., Heinrich, S., Fries, L., Antonyuk, S., Haider, C., & Salman, A. D. (2013). Adhesion mechanisms between water soluble particles. *Powder Technology, 238*, 35–49.

Dreier, W. (1991). The nuts and bolts of coating and enrobing. *Prepared Foods, 160*, 47–48.

Duri, A., George, M., Saad, M., Gastaldi, E., Ramonda, M., & Cuq, B. (2013). Adhesion properties of wheat-based particles. *Journal of Cereal Science, 58*(1), 96–103. https://doi.org/10.1016/j.jcs.2013.03.015.

Elayedath, S., & Barringer, S. A. (2002). Electrostatic powder coating of shredded cheese with antimycotic and anticaking agents. *Innovative Food Science and Emerging Technologies, 3*, 385–390.

Enggalhardjo, M., & Narsimhan, G. (2005). Adhesion of dry seasoning particles onto tortilla chip. *Food Engineering and Physical Properties, 70*(3), E215–E222. https://doi.org/10.1111/j.1365-2621.2005.tb07138.x.

Ermis, E. (2011). *Establishment of a repeatable test procedure for measuring adhesion strength of particulates in contact with surfaces*. London: University of Greenwich.

Ermis, E., Farnish, R. J., Berry, R. J., & Bradley, M. S. A. (2011). Centrifugal tester versus a novel design to measure particle adhesion strength and investigation of effect of physical characteristics (size, shape, density) of food particles on food surfaces. *Journal of Food Engineering, 104*(4), 518–524.

Griffith, E. (1991). *Cake formation in particulate systems* (p. 237). New York: VCH Publishers.

Halim, F., & Barringer, S. A. (2007). Electrostatic adhesion in food. *Journal of Electrostatics, 65*(3), 168–173.

Halim, F., & Barringer, S. A. Ã. (2015). Electrostatic adhesion in food. *Journal of Electrostatics, 65*, 168–173.

Hamaker, H. C. (1937). The London-van der Waals attraction between spherical particles. *Physica, 4*, 1058–1072.

Hu, S., Kim, T. H., Park, J. G., & Busnaina, A. A. (2010). Effect of different deposition mediums on the adhesion and removal of particles. *Journal of the Electrochemical Society, 157*(6), H662–H665. https://doi.org/10.1149/1.3377090.

Hui, Y. H. (2006). *Handbook of food products manufacturing*. Hoboken: Wiley.

Johnson, K. L., Kendall, K., & Roberts, A. D. (1971). Surface energy and the contact of elastic solids. In *Proceedings of the Royal Society A: Mathematical, Physical and Engineering Sciences*. London: Royal Society.

Kappl, M., & Butt, H. J. (2002). The colloidal probe technique and its application to adhesion force measurements. *Particle and Particle Systems Characterization, 19*, 129.

Karasu, E. N., & Ermis, E. (2019). Determination of the effect of exopolysaccharide (EPS) from lactobacillus brevis E25 on adhesion of food powders on the surfaces, using the centrifuge technique. *Journal of Food Engineering, 242*, 106–114.

Kendall, K., & Stainton, C. (2001). Adhesion and aggregation of fine particles. *Powder Technology, 121*(2–3), 223–229.

Khan, M. K. I., Schutyser, M. A. I., Schroën, K., & Boom, R. M. (2012). Electrostatic powder coating of foods - state of the art and opportunities. *Journal of Food Engineering, 111*, 1–5.

Kilcast, D., & Roberts, C. (1998). Perception and measurement of stickiness in sugar-rich foods. *Journal of Texture Studies, 29*, 81–100.

Knoll, J., Knott, S., & Nirschl, H. (2015). Characterization of the adhesion force between magnetic microscale particles and the in fluence of surface-bound protein. *Powder Technology, 283*, 163–170. https://doi.org/10.1016/j.powtec.2015.05.028.

Kobayakawa, M., Kiriyama, S., Yasuda, M., & Matsusaka, S. (2015). Microscopic analysis of particle detachment from an obliquely oscillating plate. *Chemical Engineering Science, 123*, 388–394.

Kumar, A., Staedler, T., & Jiang, X. (2013). Role of relative size of asperities and adhering particles on the adhesion force. *Journal of Colloid and Interface Science, 409*, 211–218.

LaMarche, K. R., Muzzio, F. J., Shinbrot, T., & Glasser, B. J. (2010). Granular flow and dielectrophoresis: The effect of electrostatic forces on adhesion and flow of dielectric granular materials. *Powder Technology, 199*, 180–188.

Lifshitz, E. M. (1955). The theory of molecular attractive forces between solids. *Journal of Experimental and Theoretical Physics, 2*, 94–110.

Liu, G., Li, S., & Yao, Q. (2011). A JKR-based dynamic model for the impact of micro-particle with a flat surface. *Powder Technology, 207*, 215–223.

Lusas, E., & Lloyd, R. (2001). *Snack Foods Processing (Google eBook)* (p. 639). Boca Raton: CRC Press. Available at: http://books.google.com/books?id=W_5wlzckPkMC&pgis=1.

Megias-Alguacil, D., & Gauckler, L. J. (2009). Capillary forces between two solid spheres linked by a concave liquid bridge: Regions of existence and forces mapping. *AIChE Journal, 55*, 1103–1109.

Miller, M. J., & Barringer, S. A. (2002). Effect of sodium chloride particle size and shape on nonelectrostatic and electrostatic coating of popcorn. *Journal of Food Science, 67*(1), 198–201.

Niman, C. E. (2000). In search of the perfect salt for topping snack foods. *Cereal Foods World, 45*(10), 466–469.

Nussinovitch, A. (2017). Adhesion in foods. In *Fundamental principles and applications*. West Sussex: Wiley.

Packham, D. E. (2003). Surface energy, surface topography and adhesion. *International Journal of Adhesion and Adhesives, 23*, 437–448.

Payam, A. F., & Fathipour, M. (2011). A capillary force model for interactions between two spheres. *Particuology, 9*, 381–386.

Peleg, M. (1977). Flowability of food powders and methods for its evaluation — A review. *Journal of Food Process Engineering, 1*(4), 303–328.

Petean, P. G. C., & Aguiar, M. L. (2015). Determining the adhesion force between particles and rough surfaces. *Powder Technology, 274*, 67–76. Available at: https://doi.org/10.1016/j.powtec.2014.12.047.

Pietsch, W. (2003). An interdisciplinary approach to size enlargement by agglomeration. *Powder Technology, 130*, 8–13.

Prasad, L. K., McGinity, J. W., & Williams, R. O. (2016). Electrostatic powder coating: Principles and pharmaceutical applications. *International Journal of Pharmaceutics, 505*(1–2), 289–302. Available at: https://linkinghub.elsevier.com/retrieve/pii/S037851731630299X.

Rabinovich, Y. I., Esayanur, M. S., & Moudgil, B. M. (2005). Capillary forces between two spheres with a fixed volume liquid bridge: Theory and experiment. *Langmuir, 21*, 10992–10997.

Riaz, M. N. (2015). Snack foods: rocessing. In *Encyclopedia of food grains* (2nd ed.). San Diego: Elsevier Science.

Ripperger, S., & Hein, K. (2008). Measurement of adhesion forces in air with the vibration method. *China Particuology, 3*(1), 3–9.

Rumpf, H., & Knepper, W. (1962). The strength of granules and agglomerates. In *International symposium on agglomeration* (pp. 379–418). Rugby: Institution of Chemical Engineers.

Salazar-Banda, G. R., Felicetti, M. A., Gonçalves, J. A. S., Coury, J. R., & Aguiar, M. L. (2007). Determination of the adhesion force between particles and a flat surface, using the centrifuge technique. *Powder Technology, 173*(2), 107–117.

Schubert, H. (1987). Food particle technology. Part I: Properties of particles and particulate food systems. *Journal of Food Engineering, 6*(1), 1–32.

Shukla, N., & Henthorn, K. H. (2009). Effect of relative particle size on large particle detachment from a microchannel. *Microfluidics and Nanofluidics, 6*(4), 521–527.

Simons, S. J. R. (2007). Liquid bridges in granules. In *Granulation* (Vol. 11). Amsterdam: Elsevier. (Handbook of Powder Technology).

Suganya, A., Shanmugvelayutham, G., & Hidalgo-Carrillo, J. (2018). Plasma surface modified polystyrene and grafted with chitosan coating for improving the shelf lifetime of postharvest grapes. *Plasma Chemistry and Plasma Processing, 38*(5), 1151–1168. Available at: http://link.springer.com/10.1007/s11090-018-9908-0.

Sumawi, H., & Barringer, S. A. (2005). Positive vs. negative electrostatic coating using food powders. *Journal of Electrostatics, 63*, 815–821.

Takeuchi, M. (2006). Adhesion forces of charged particles. *Chemical Engineering Science, 61*, 2279–2289.

Teunou, E., Fitzpatrick, J. J., & Synnott, E. C. (1999). Characterization offood powder flowability. *Journal of Food Engineering, 39*(1), 31–37.

Vahdat, A. S., Azizi, S., & Cetinkaya, C. (2013). Nonlinear dynamics of adhesive micro-spherical particles on vibrating substrates. *Journal of Adhesion Science and Technology, 27*(15), 1712–1726.

Wang, J. P., Gallo, E., François, B., Gabrieli, F., & Lambert, P. (2017). Capillary force and rupture of funicular liquid bridges between three spherical bodies. *Powder Technology, 305*, 89–98.

Wanka, S., Kappl, M., Wolkenhauer, M., & Butt, H. (2013). Measuring adhesion forces in powder collectives by inertial detachment. *Langmuir, 29*, 16075–16083.

Wong, D. C. Y., Adams, M. J., Seville, J. P. K., & Fryer, P. J. (2005). A computational model of flavour deposition onto food surfaces. *Food and Bioproducts Processing, 83*, 99–106.

Yousuf, S., & Barringer, S. A. (2007). Modeling nonelectrostatic and electrostatic powder coating. *Journal of Food Engineering, 83*, 550–561.

Zafar, U., Hare, C., Hassanpour, A., & Ghadiri, M. (2014). Drop test: A new method to measure the particle adhesion force. *Powder Technology, 264*, 236–241. https://doi.org/10.1016/j.powtec.2014.04.022.

Chapter 4
Characterization of the Caking Behaviour of Food Powders

John J. Fitzpatrick

4.1 Introduction

There are many different powder caking mechanisms, including van der Waals attractions, electrostatic attractions, liquid bridging and solid bridging (Zafar et al. 2017). Van der Waals attractions are caused by induced dipoles on powder surfaces, and the strength of any cakes formed depend on how compacted the powder is. They can become very strong when highly compacted, such as in tabletting. Liquid bridging in food powders is usually associated with liquid being presented in the interstitial spaces between the powder particles giving rise to capillary forces which hold the particles together. Liquid bridging may give rise to weak cakes and lumping, while solid bridge formation results in much stronger cakes. Solid bridge formation in food powders usually occurs due to exposing the powder to a relative humidity cycle or due to sinter bridge formation in a powder containing significant amorphous content.

Humidity cycling is where water soluble powders are first exposed to high relative humidity, which is followed by exposure to low relative humidity. Water bridges between particles containing dissolved solids form during high humidity exposure. During subsequent low humidity exposure, water vaporises leaving the solids behind which may be sufficient to form solid bridges between the particles. For water-soluble crystalline powders, the powder will sorb much water when exposed to a relative humidity greater that its deliquescence relative humidity giving rise to water bridges containing dissolved solids. Initiation of solid bridge formation is triggered by exposing the powder to a lower relative humidity that is below its efflorescence relative humidity (Carpin et al. 2016), where sufficient water has vaporised

J. J. Fitzpatrick (✉)
Process and Chemical Engineering, School of Engineering, University College Cork, Cork, Ireland
e-mail: j.fitzpatrick@ucc.ie

E. Ermiş (ed.), *Food Powders Properties and Characterization*, Food Engineering Series, https://doi.org/10.1007/978-3-030-48908-3_4

to produce a supersaturated solution, which enables crystallization and the potential for solid bridge formation.

Sinter bridge formation in food powders is often due to the presence of amorphous carbohydrates, such as sugars and shorter chain sugar polymers. These amorphous components may become sticky (Aguilera et al. 1995; Descamps et al. 2013) if the powder temperature is elevated to its glass transition temperature (Tg) and above, or if Tg is reduced to the powder temperature or below. At these temperatures where the powder temperature is at or above Tg, the powder develops viscous behaviour that may give rise to stickiness and caking. The powder particles may form sinter bridges by flowing into each other, which can give rise to very strong cakes (Palzer 2005). Water content has a big influence on Tg because water sorption causes a reduction in Tg. Consequently, if an amorphous powder sorbs water that causes it Tg to decrease below its temperature, then it may become sticky and form sinter bridges resulting in strong cakes. Thus, the caking of amorphous food powders is strongly dependent on the relative humidity of the surrounding air and temperature.

There are many different techniques for measuring powder cake strength including shear test techniques (Descamps et al. 2013) uniaxial compression testing (Weigl et al. 2006), force-displacement test to extrude a plug from a caked powder (Fitzpatrick 2008, 2010), penetration testing (Chen et al. 2017), powder rheometer (Chávez Montes et al. 2011) and blow tester (Paterson et al. 2005). Zafar et al. (2017) provides a good summary of various techniques to measure powder cake strength in their review paper on powder caking. Visualisation techniques are also useful for exploring caking mechanisms, such as the use of light microscopes and electron microscopes (Feeney and Fitzpatrick 2011). Visual assessment of the bulk powder allows identification of changes in physical characteristics of the cake, such as hardness, wetness and powder shrinkage during caking. Visualisation techniques may be used to complement the results from cake strength testers. Descamps and Palzer (2007) showed that maltodextrin DE 21 formed a very strong cake when the powder was exposed to 76% relativity humidity (25 °C), which caused its glass transition temperature to decrease below the powder temperature. They used a light microscope to provide photographs of two maltodextrin particles in contact when exposed to the same condition over time. The photographs showed that the powder particles transitioned from their glassy state into a viscoelastic state which caused the two particles to flow forming a sinter bridge between the particles, which caused the very strong cake. Saragoni et al. (2007) applied videomicroscopy to continuously monitor changes in the projected area of individual coffee powder particles over time. They defined a caking index based on the change in projected area and applied this as a first approximation for predicting the caking tendency of amorphous particles.

In this chapter, powder caking is characterized using a force-displacement tester to measure cake strength coupled with powder water content measurement, and visualisation of the bulk powder and two particles in contact. This is based on work carried out by the author. Section 4.2 describes the techniques used and Sect. 4.3 provides examples of applying the techniques to characterising the humidity caking of amorphous maltodextrin and crystalline NaCl powders, and binary mixes of maltodextrin with either NaCl or paprika.

4.2 Physical and Visual Techniques for Characterising Caking Behaviour

This section describes the physical and visual techniques used by the author for characterising the caking behaviour of food powders.

4.2.1 *Physical Techniques*

4.2.1.1 Force Displacement Cake Strength Tester

Fitzpatrick et al. (2008) and Fitzpatrick et al. (2010) describe an empirical tester for quantitatively measuring an index of the cake strength of a powder. It consists of placing a mass of powder into a cylindrical aluminium dish which has a circular hole exactly in the centre of the dish as illustrated in Fig. 4.1a. A layer of aluminium foil is first taped to the bottom of the dish to cover the hole, preventing powder falling through the hole. The powder is spread out across the dish using a leveler to produce a consistent powder layer with a fixed thickness of 7 mm. The powder in the dish is then exposed to a constant relative humidity. Saturated LiCl and NaCl

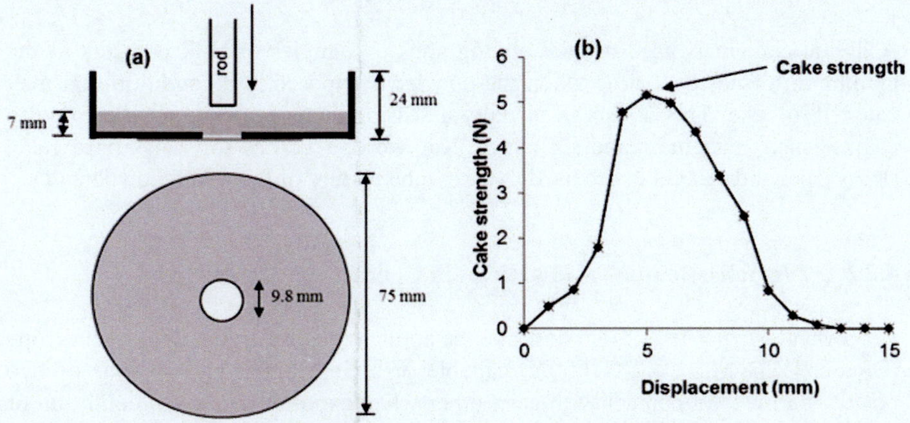

Fig. 4.1 (**a**) Aluminium dish used in force–displacement cake strength determination (with central hole and foil cover underneath to prevent powder falling through (**b**) Force vs. displacement for the cake strength tester on a skim milk powder sample exposed to 20 °C, 100% relative humidity for 20 h

solutions are used to obtain constant relative humidities of 11% and 76%, respectively, and pure water to obtain a relative humidity of 100%.

Following exposure for a specified time, the foil is removed from beneath the aluminium dish and the dish is centred below a rod attached to a TA Texture Analyser (HDplus). The rod, with a diameter of 5 mm, is initially moved downwards to just above the top of the powder surface. The rod is then moved downward through the powder at a constant speed of 30 mm/min and the force versus displacement is measured using the TA Texture Analyser as the rod moves through the cake. As the rod engages the surface of the cake, the force increases to a maximum and then gradually reduces as the rod continues to push a plug of caked powder through the hole in the centre of the dish. The peak force measured is defined as the cake strength, as illustrated in Fig. 4.1b. Measured strengths less than the 1 N are considered as not caked, as the powder typically has not caked and just has sufficient strength to span across the hole at the bottom of the dish.

4.2.1.2 Water Content Measurement

Water content measurements are performed for the original powders and for samples of powder taken from the aluminium dishes after cake strength testing. The water content is evaluated by weighing a mass of powder sample before and after drying in an oven at 104 °C for 24 h.

4.2.2 Visual Techniques

4.2.2.1 Visualisation of Bulk Powder

A digital camera is used to take photographs of samples of bulk powders in the aluminium dishes over time when the powder is exposed to a condition that may cause it to cake. This assists in visually assessing changes in the powders due to caking, such as wetness and shrinkage. Key words (such as soft cake, hard cake, sticky cake and wet cake) are used to assess the texture of the caked powders.

4.2.2.2 Visualisation of Two Particles in Contact

Feeney and Fitzpatrick (2011) describe the application of a digital light microscope (Bresser LCD Micro 40X-1600X) capable of taking digital photographs of two powder particles in contact with each other when exposed to a given condition of relative humidity and temperature, as illustrated in Fig. 4.2. Two powder particles are placed in contact with each other inside a transparent Petri dish. To produce a desired relative humidity, a small open plastic dish containing a saturated salt solution (or pure water) is placed inside the Petri dish, which is then sealed using a

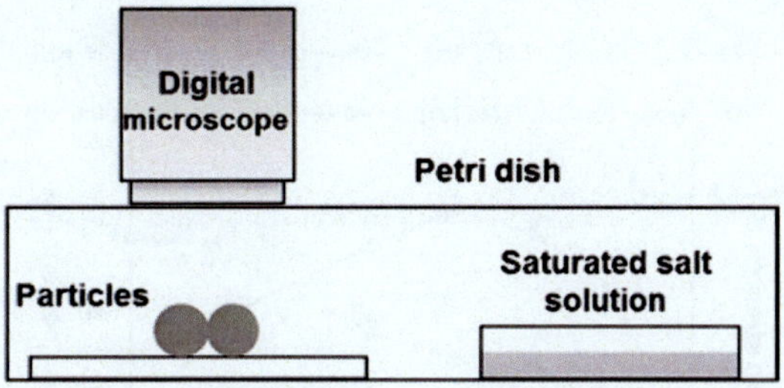

Fig. 4.2 Visual assessment of the caking behaviour of two particles in contact using a digital microscope

parafilm tape to maintain the constant relative humidity. Photographs are taken of the particles over time using the microscope to visually observe any changes occurring with the particles, such as water sorption and dissolution, particle deformation and formation of sinter bridges.

4.3 Examples of the Application of the Physical and Visual Techniques for Assessing the Caking Behaviour of Food Powders

This section provides a number of examples of the characterisation of powder caking using the techniques presented in Sect. 4.2. These examples are presenting some of the major results obtained from work conducted by the author (Fitzpatrick et al. 2010; Fitzpatrick et al. 2017), which focus on the humidity caking of individual food powders and food powder binary mixes.

4.3.1 Humidity Caking of NaCl and Maltodextrin DE21

The cake strength and water content of the NaCl powder when exposed to 76% and 100% relative humidity (RH) over a 7 day duration (25 °C) are presented in Fig. 4.3a. At 76% RH, the NaCl powder did not cake over the 7 day exposure time, having no or very low strengths at less than 1 N. The NaCl powder sorbed little water with its water content increasing from 0.1 to 0.7%. Photographs of two particles in contact (Fig. 4.3b) shows that the particles did not change their shape over a 60 min exposure time. Longer exposure times up to 7 days also showed no change in particle shape, however a small amount of liquid bridging was observed at the contact point between the particles. At 100% RH exposure, the NaCl powder progressively sorbed water, being nearly 15% at day 7. The initial water sorption resulted in an increase of cake strength up to around 2.5 N at day 3 before decreasing back to around 1 N due to the progressive increase in water content. The reason for the water sorption at 100% RH is simply because the NaCl powder is exposed to a RH that is greater than its deliquescence RH, which is around 76% for NaCl. If a water soluble crystal,

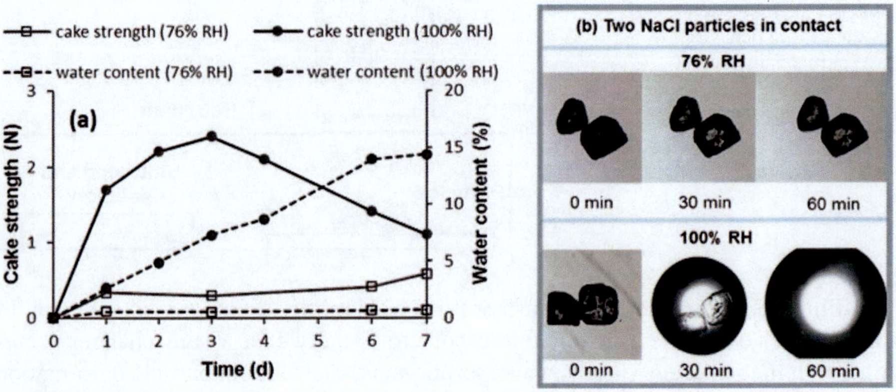

Fig. 4.3 Humidity caking of NaCl at 76% and 100% RH: (**a**) Evolution of powder cake strength and water content over 7 days (25 °C), (**b**) photographs of two powder particles in contact over time (20 °C)

such as NaCl, is exposed to a RH that is higher than its deliquescence RH, it will start to sorb water from the surrounding atmosphere and will eventually dissolve in the water sorbed over time (Hartmann and Palzer 2011). This is nicely shown by photographs of the two particles in contact (Fig. 4.3b) where the two particles sorb much water and eventually dissolve over a 60 min exposure time.

The effect of exposing the maltodextrin powder to 76% and 100% RH (25 °C) on powder cake strength and water content over 7 days is illustrated in Fig. 4.4a. The powder slowly but readily sorbed water over time. At 76% RH, the powder did not cake after 1-day exposure, but there was a major increase in cake strength to around 175 N at day 4. At 7 day exposure, the powder was visually transformed from a powder into a hard shiny marble like material (Fig. 4.4b) that displayed major shrinkage. This caking behavior was caused by sinter bridging due to a reduction in its glass transition temperature induced by water sorption. Spray-dried maltodextrin DE21 is an amorphous material and has an onset glass transition temperature of 25 °C when the wet basis water content is 10.7%. The water content of maltodextrin was about 11% at the end of 2-day exposure, thus temperature of the powder was greater than its glass transition temperature. Visualisation of the effect of exposing 2 maltodextrin particles in contact to 76% RH (20 °C) is presented in Fig. 4.4c. This shows the formation of a sinter bridge over time between the maltodextrin particles. By 240 min exposure time, the particles have totally flowed into each other forming a round shape. At this stage, the particles have transformed from a glassy material into a more liquid like material. At the higher RH of 100% RH, the maltodextrin powder sorbed more water more rapidly, and consequently caked more rapidly, as illustrated in Fig. 4.4a. However, its water sorption behaviour resulted in a major reduction in cake strength to 20 N at day 3, which continued to decrease to a very weak cake strength of 2 N at the end of day 5, as the cake transformed itself from a hard solid to a more viscoelastic material. Two maltodextrin particles in contact

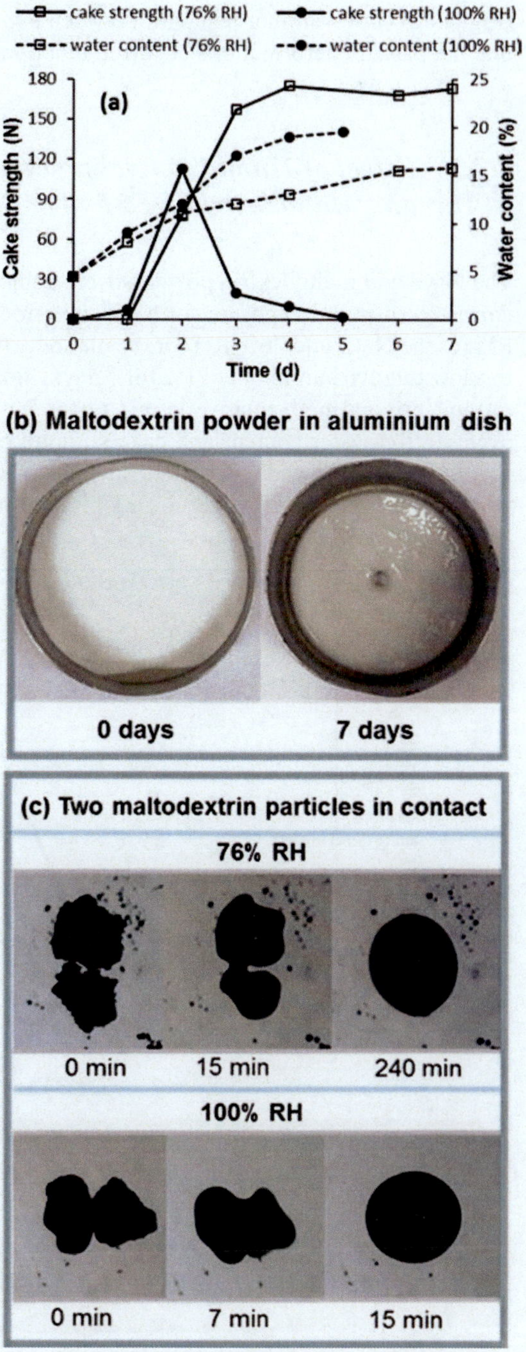

Fig. 4.4 Humidity caking of maltodextrin powder at 76% and 100% RH: (**a**) Evolution of powder cake strength and water content over 7 days (25 °C), (**b**) photographs of powder in aluminium dish at 0 and 7 day exposure times at 76% RH, (**c**) photographs of two powder particles in contact over time (20 °C)

(Fig. 4.4c) shows similar behavior between exposure at 76% and 100% RH, except that the particle deformations occurred much more rapidly at 100% RH.

4.3.2 *Effect of Humidity Cycling on the Caking of NaCl and Maltodextrin DE21*

The NaCl and maltodextrin powders were subjected to a humidity cycle. This cycle consisted of initial exposure for a 2-day period to a high relative humidity (100% RH for the NaCl and 76% RH for the maltodextrin). This was followed by exposure to a low relative humidity at 11% for 2 days. Finally, the powders were then exposed to their original high relative humidity for 2 days. The influence of the humidity cycle on the water content and cake strength of the NaCl powder is illustrated in Fig. 4.5a. This showed a classic behaviour for a water soluble crystalline material, where the powder sorbs water during high humidity exposure, which causes water

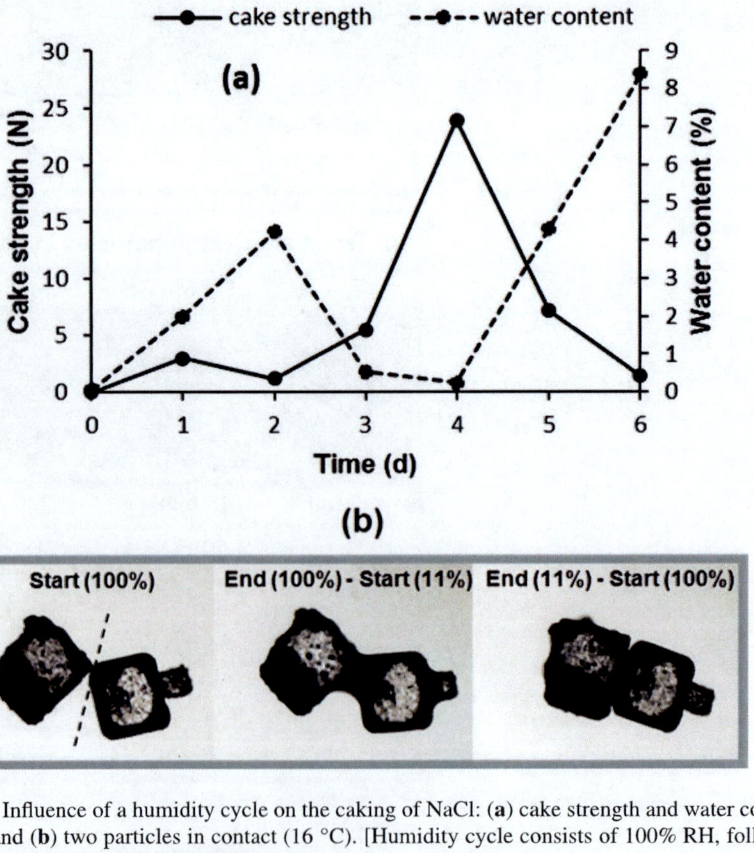

Fig. 4.5 Influence of a humidity cycle on the caking of NaCl: (**a**) cake strength and water content (25 °C) and (**b**) two particles in contact (16 °C). [Humidity cycle consists of 100% RH, followed by 11% RH, followed by 100% RH]

bridges and surface dissolution into these bridges, which dry out to form solid bridges during low humidity exposure. The water content of the NaCl powder increased to about 4% over the first 2-day exposure to the high 100% RH. The NaCl did cake, however it was a very weak cake at around 1–3 N. The water content decreased rapidly when exposed to the low 11% RH down to about 0.25% after two-day exposure. The vaporisation of water caused a big increase in cake strength, increasing to 24 N at the second day of exposure to 11% RH. This is due to solid bridge formation caused by recrystallisation of dissolved NaCl in the liquid bridges between the crystals. The cake sorbed water once again when it was re-exposed to the high relative humidity, resulting in the cake strength decreasing back to around 1 N as the crystal bridges were dissolved due to the formation of liquid bridges between the NaCl crystals.

Exposing two NaCl particles to the humidity cycle is illustrated in Figure 4.5b. As already seen (Fig. 4.3b), the particles readily sorbed water when exposed to 100% RH and a liquid bridge was formed. During the low humidity exposure, the water was vaporised and there appeared to be a solid bridge between the particles, as illustrated in Fig. 4.5b. The presence of a solid bridge was identified by probing the particles with a pin. The liquid bridge reformed during re-exposure to the high RH and the particles eventually dissolved as illustrated in Fig. 4.3b.

The influence of the humidity cycle on the water content and cake strength of maltodextrin is illustrated in Fig. 4.6a. The water content and cake strength of the maltodextrin rose to 11.5% and 78 N, respectively, after two-day exposure to 76% RH. Both the water content and the cake strength of the maltodextrin decreased during subsequent 2-day exposure to the low relative humidity environment of 11% RH. The water content decreased to 7% after 2 day exposure. As mentioned previously, the maltodextrin powder has a Tg of 25 °C at 10.7%, thus the sinter bridges were in the glassy state. Initially, it was expected that the presence of glassy sinter bridges would lead to even stronger cakes, however the cakes were weaker, as illustrated in Fig. 4.6b. The cake became more brittle as the it lost water and became glassy, it lost its toughness and failed more easily, resulting in a decrease in cake strength. However, it should be noted that the low cake strength was 13 N which is still a cake with appreciable strength. Re-exposing the maltodextrin to 76% RH caused an increase in both powder cake strength and water content. After 2 day re-exposure to 76% RH, the water content was high enough for the glass transition temperature to drop well below 25 °C. Consequently, the cake strength increased to around 100 N because the sinter bridges were no longer glassy.

Exposing two maltodextrin particles to the humidity cycle is illustrated in Fig. 4.6b. During high humidity exposure (76% RH), the particles readily changed shape, flowed and a bridge was formed between the particles. At the end of the low humidity exposure, the size of the bridge and size and shape of the particles were similar to that at the beginning of the low humidity exposure. This suggests that the bridge established between the particles during high humidity exposure was due to sintering rather than the establishment of a water bridge. During re-exposure to the high humidity, the particles sorbed water and flowed into each other because of their viscoelastic behavior.

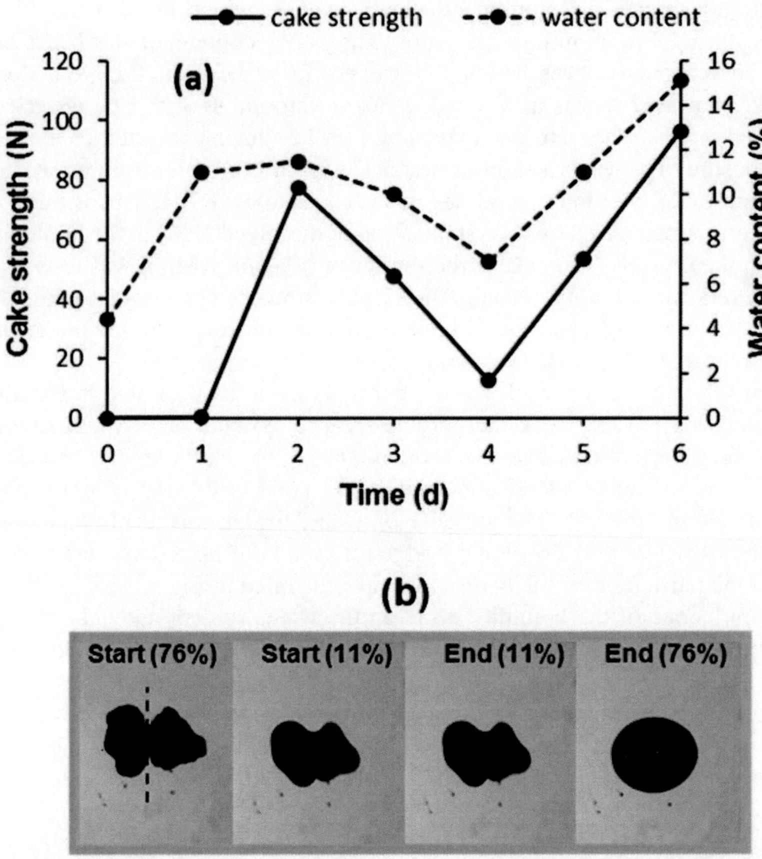

Fig. 4.6 Influence of a humidity cycle on the caking of maltodextrin: (**a**) cake strength and water content (25 °C) and (**b**) two particles in contact (16 °C). [Humidity cycle consists of 76% RH, followed by 11% RH, followed by 76% RH]

4.3.3 Humidity Caking of Binary Mixes Containing Maltodextrin DE21

Binary powder mixes were formulated consisting of maltodextrin DE21 with either paprika or NaCl. The mixes were formulated at compositions containing 40% and 80% maltodextrin (MD). The mixes were exposed to a 76% RH atmosphere. The cake strength and water content were measured over time, along with visual assessment of the bulk. Complementary visual assessment of two particles in contact were also undertaken.

4.3.3.1 MD/Paprika Mixes

The cake strength and water content of the paprika powder when exposed to 76% RH over a 28 day exposure period is presented in Fig. 4.7. The paprika readily sorbed water with its water content increasing from about 10% up to about 19%. The cake strength was very low reaching only about 0.2 N, which means that the paprika powder remains essentially uncaked over the 28 day exposure time.

The effect of exposure time on the cake strength of MD/paprika mixes and individual powders is presented in Fig. 4.8a. For the 40% MD mix, the cake strength is much weaker than that of the 100% MD, displaying a cake strength of only 2 N at 28 day exposure. This is most likely due to insufficient MD in the mix to cause sufficient sinter bridging, in particular between the MD particles themselves. For the 80% MD mix, much stronger cakes were formed, which initially increased to about 70 N at 10 days, but then decreased to 30 N at 14 days and remained at this value at 28 days. In relation to the evolution of water contents over time, Fig. 4.8b shows that the paprika mixes had values in between 100% paprika and 100% MD.

Visualisation of the caking behaviour of the MD/paprika bulk powder mixes is presented in Fig. 4.9a at 3, 10 and 28 day exposures for the 40% and 80% MD mixes (76% RH, 20 °C). There were changes in textural properties (and colour) for both mixes over time, especially the mix with 80% MD. These showed progressive powder shrinkage, which is most likely due to glass transition induced sinter bridging caused by deformation of the MD particles. Visualisation of a MD particle and a paprika particle in contact over time was undertaken when the particles were exposed to 76% RH (20 °C). The MD particle deformed over time and the paprika did not, as illustrated in Fig. 4.9b, and a sinter bridge was formed due to the MD

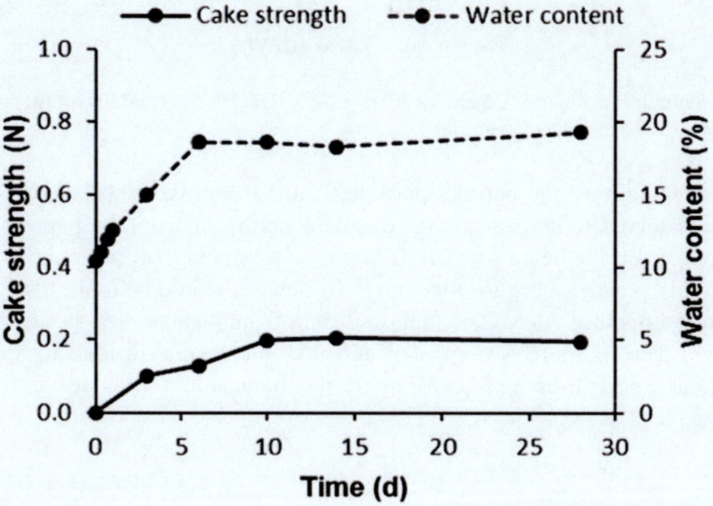

Fig. 4.7 Effect of exposure to 76% RH (20 °C) on cake strength and water content of paprika powder over 28 days

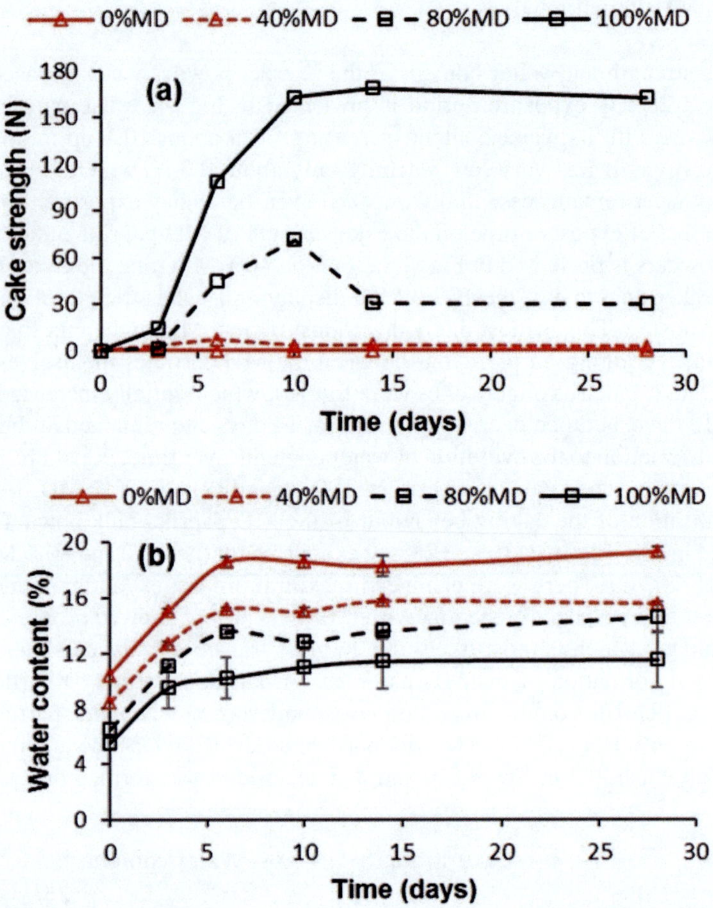

Fig. 4.8 Humidity caking of MD/paprika mixes at 76% RH (20 °C)–Evolution of (**a**) powder cake strength and (**b**) water content over time

particle flowing onto the paprika particle. In other similar trials, the MD particle was also observed to deform but instead detached itself from the paprika particle, with no sinter bridge being formed. It might be expected to observe more sintering within a bulk powder, because the weight of powder would be more likely to force the particles together. As MD content increases from 40% to 80% in the bulk powder mixes, there is more MD present and thus more sinter bridge formation and stronger cakes, due to more MD/MD particle contacts in particular contributing to stronger cakes.

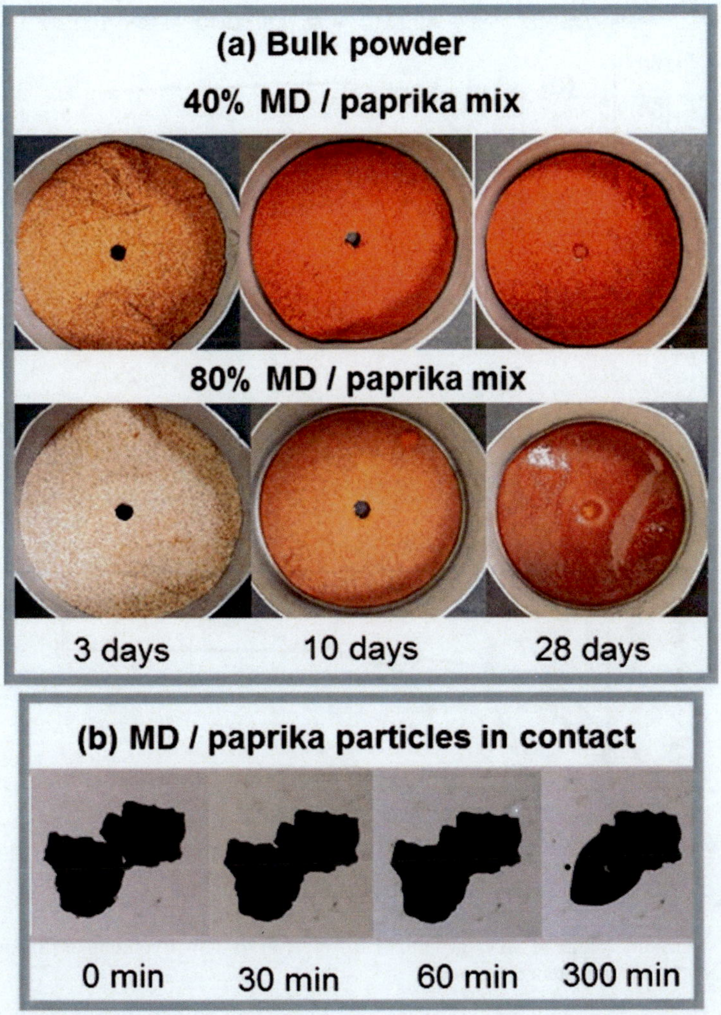

Fig. 4.9 Humidity caking of MD / paprika mixes at 76% RH (20 °C)—Visualisation of (**a**) bulk powder and (**b**) two particles in contact over time

4.3.3.2 MD/NaCl Mixes

Figure 4.10a illustrates the evolution of cake strength of the MD/NaCl mixes and the individual powders, when exposed to 76% RH. There is a similar evolution in cake strength of the mixes and the100% MD powder over the initial 3 days, after which the cake strength shows a major reduction. For both mixes, the cake strength of both mixes was reduced to a very low strength of about 0.5 N at 6 day exposure, and remained at around 1 N for the rest of the 28 day exposure.

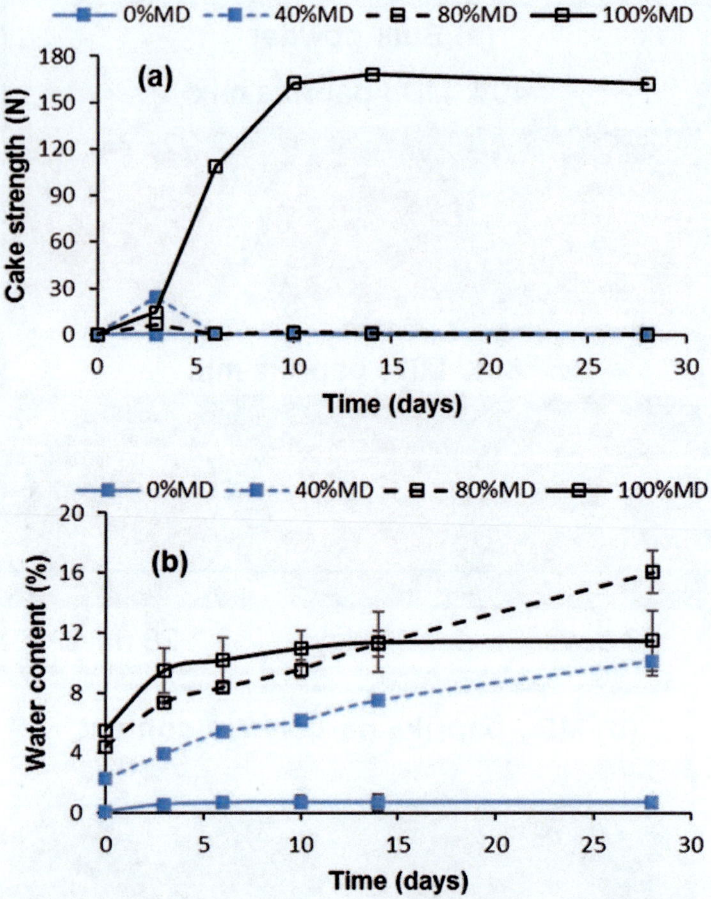

Fig. 4.10 Humidity caking of MD/NaCl mixes at 76% RH (20 °C)—Evolution of (**a**) powder cake strength and (**b**) water content over time

The evolution of water contents are presented in Fig. 4.10b. The water content of the individual MD powder increased more rapidly over the first 3 days and then increased more gradually to around 11.5% at day 14. The water content of the NaCl was very low throughout the 28 day exposure at around 0.7%. The water contents for the MD/NaCl mixes continuously increased over time with that of the 40% MD mix approximating that of the 100% MD at 28 day exposure. For the 80% MD mix, Fig. 4.10b shows that its water content surpassed that of the 100% MD mix at around 14 days and was much greater at 28 day exposure. From this, the weak caking behaviour of the MD/NaCl mixes is most likely caused by their water sorption behaviour.

The visual observation of the bulk MD/NaCl mixtures was undertaken and this substantiates their cake strength behaviour. Some shrinkage was observed after 3 days for the 40% MD/NaCl mixes and a soft, deformable cake was formed. The cake became wetter after longer exposure times and also became stickier but was easily broken. The 80% looked similar to the 40% MD/NaCl mixes after 3 days with some shrinkage being evident, however, there was a noticeable difference after 28 days, as illustrated in Fig. 4.11a. The bulk was shiny, easily deformable, wet and starting to flow, giving rise to a very weak cake.

Figure 4.11b shows photographs of the MD and NaCl particles in contact at 76% RH. The MD particle deformed readily (as already observed in Fig. 4.4b) and a sinter bridge was formed due to the MD flowing onto the NaCl particle. This may explain the initial evolution of cake strength over the first 3 days. The flow of MD continued over time and the NaCl particle appeared to gradually dissolve, as illustrated in Fig. 4.11b. This is possibly being caused by the deliquescence lowering phenomena; in particular the lowering of the deliquescence RH of NaCl. Salameh et al. (2006) showed that deliquescence lowering can occur when two chemically different deliquescent food ingredient powders are contacted, such as in a powder mix. This could lead to a mix sorbing much water even when the exposure RH is less than the deliquescence RH for each of the individual powders. For NaCl, its deliquescence RH is around 76%, which is the same as the exposure RH. Contact with MD may have lowered its deliquescence RH, resulting in the continuous water sorption by the MD/NaCl mixes over time (Fig. 4.10b), causing the formation of the wet weak cakes observed.

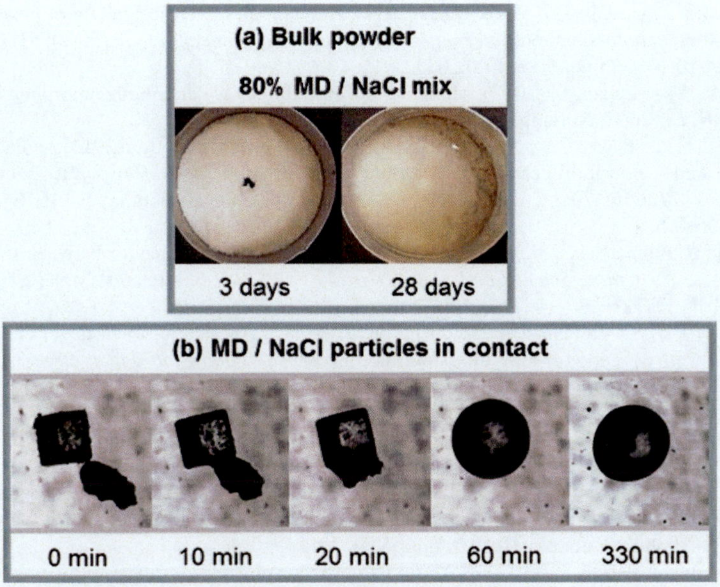

Fig. 4.11 Visualisation of humidity caking of MD/NaCl mixes at 76% RH (20 °C): (**a**) bulk powder and (**b**) two particles in contact over time

4.4 Conclusions

A combination of physical and visual techniques were applied to assess the caking behaviour of food powders when exposed to humidity caking conditions. The physical techniques consisted of cake strength and water content measurements which allowed quantitative assessment of caking behaviour. The visual techniques consisted of observation of bulk powder caking behaviour (identification of powder shrinkage and changes in physical characteristics of the cake), and the use of a microscope to visually assess two particles in contact in terms of water uptake and dissolution, particle deformation and sinter bridge formation. The combination of techniques complement each other in providing a better description and understanding of the caking behaviour of food powders.

References

Aguilera, J. M., Valle, J. M., & Ka, M. (1995). Caking phenomena in amorphous food powders. *Trends in Food Science & Technology, 6*, 149–155.

Carpin, M., Bertelsen, H., Bech, J. K., Jeantet, R., Risbo, J., & Schuck, P. (2016). Caking of lactose: A critical review. *Trends in Food Science & Technology, 53*, 1–12. https://doi.org/10.1016/j.tifs.2016.04.002.

Chávez Montes, E., Ardila Santamaría, N., Gumy, J.-C., & Marchal, P. (2011). Moisture-induced caking of beverage powders. *Journal of the Science of Food and Agriculture, 91*(14), 2582–2586. https://doi.org/10.1002/jsfa.4496.

Chen, Q., Zafar, U., Ghadiri, M., & Bi, J. (2017). Assessment of surface caking of powders using the ball indentation method. *International Journal of Pharmaceutics, 521*(1–2), 61–68. https://doi.org/10.1016/j.ijpharm.2017.02.033.

Descamps, N., & Palzer, S. (2007). Modeling the sintering of water soluble amorphous particles. In *PARTEC 2007*. Nuremberg: Nürnberg Messe.

Descamps, N., Palzer, S., Roos, Y. H., & Fitzpatrick, J. J. (2013). Glass transition and flowability/caking behaviour of maltodextrin DE 21. *Journal of Food Engineering, 119*(4), 809–813. https://doi.org/10.1016/j.jfoodeng.2013.06.045.

Feeney, J., & Fitzpatrick, J. J. (2011). Visualization of the caking behavior between two powder particles. *Particulate Science and Technology, 29*(5), 397–406. https://doi.org/10.1080/02726351.2010.503324.

Fitzpatrick, J. J., O'Callaghan, E., & O'Flynn, J. (2008). Application of a novel cake strength tester for investigating caking of skim milk powder. *Food and Bioproducts Processing, 86*(3), 198–203. https://doi.org/10.1016/j.fbp.2007.10.009.

Fitzpatrick, J. J., Descamps, N., O'Meara, K., Jones, C., Walsh, D., & Spitere, M. (2010). Comparing the caking behaviours of skim milk powder, amorphous maltodextrin and crystalline common salt. *Powder Technology, 204*(1), 131–137. https://doi.org/10.1016/j.powtec.2010.07.029.

Fitzpatrick, J. J., O'Connor, J., Cudmore, M., & Dos Santos, D. (2017). Caking behaviour of food powder binary mixes containing sticky and non-sticky powders. *Journal of Food Engineering, 204*, 73–79. https://doi.org/10.1016/j.jfoodeng.2017.02.021.

Hartmann, M., & Palzer, S. (2011). Caking of amorphous powders — material aspects, modelling and applications. *Powder Technology, 206*(1–2), 112–121. https://doi.org/10.1016/j.powtec.2010.04.014.

Palzer, S. (2005). The effect of glass transition on the desired and undesired agglomeration of amorphous food powders. *Chemical Engineering Science, 60*(14), 3959–3968. https://doi.org/10.1016/j.ces.2005.02.015.

Paterson, A. H. J., Brooks, G. F., Bronlund, J. E., & Foster, K. D. (2005). Development of stickiness in amorphous lactose at constant T−Tg levels. *International Dairy Journal, 15*(5), 513–519. https://doi.org/10.1016/j.idairyj.2004.08.012.

Salameh, A. K., Mauer, L. J., & Taylor, L. S. (2006). Deliquescence Lowering in Food Ingredient Mixtures. *Journal of Food Science, 71*(1), E10–E16. https://doi.org/10.1016/j.foodchem.2006.11.029.

Saragoni, P., Aguilera, J. M., & Bouchon, P. (2007). Changes in particles of coffee powder and extensions to caking. *Food Chemistry, 104*(1), 122–126. https://doi.org/10.1016/j.foodchem.2006.11.029.

Weigl, B., Pengiran, Y., Feise, H. J., Röck, M., & Janssen, R. (2006). Comparative testing of powder caking. *Chemical Engineering & Technology, 29*(6), 686–690. https://doi.org/10.1002/ceat.200600083.

Zafar, U., Vivacqua, V., Calvert, G., Ghadiri, M., & Cleaver, J. A. S. (2017). A review of bulk powder caking. *Powder Technology, 313*, 389–401.

Chapter 5
Characterisation of the Rehydration Behaviour of Food Powders

John J. Fitzpatrick, Junfu Ji, and Song Miao

5.1 Introduction

Most food powders are eventually rehydrated before use (Schubert 1993). Consequently, the ability of a powder to rehydrate is an essential powder functionality, as the industrial users and home consumers do not need encountering problems during rehydration. Food powders may inherently have different rehydration abilities (Barbosa-Canovas et al. 2005), thus it is important to be able to characterise their rehydration behaviour.

Powder rehydration is usually broken down into a number of steps. These include wetting, sinking, dispersion, swelling, disintegration and solubilisation or dissolution (Freudig et al. 1999; Goalard et al. 2006; Forny et al. 2011; Mitchell et al. 2015). Wetting involves the initial contact between the powder and water, where water penetrates into the voids between the powder particles and wets their surfaces. Sinking refers to powder submerging below the water surface and into the bulk water. Dispersion refers to the transport of wetted powder particulates throughout the bulk water. This may include the disaggregation of agglomerates into their primary powder particles. Solubilisation involves the dissolution of powder to produce a solution. Powder particles may swell whereby water penetrates into the particles

J. J. Fitzpatrick (✉)
Process and Chemical Engineering, School of Engineering, University College Cork, Cork, Ireland
e-mail: j.fitzpatrick@ucc.ie

J. Ji
College of Food Science and Nutritional Engineering, China Agricultural University, Beijing, China
e-mail: junfu.ji@cau.edu.cn

S. Miao
Teagasc Food Research Centre, Moorepark, Fermoy, Co. Cork, Ireland
e-mail: Song.Miao@teagasc.ie

E. Ermiş (ed.), *Food Powders Properties and Characterization*, Food Engineering Series, https://doi.org/10.1007/978-3-030-48908-3_5

causing them to expand in size. These steps may occur together during some periods of time, such as powder particles dissolving as they disperse. Consequently, it is difficult to measure and investigate these steps in isolation. In stirred systems, the rehydration process may be broken down into the two key steps of wetting and solubilisation.

There are a number of experimental methods reported that can be applied to assess the rehydration behaviour of powders. For wetting, these include the International Dairy Federation standard method for measuring wetting time (Anon 1979), dynamic contact angle measurement (Dupas et al. 2013), Washburn method (Washburn 1921) and the turbidity method (Gaiani et al. 2009). For solubilisation ability, methods include the measurements of particle size, dissolved solids and turbidity over time in a specified agitation system (Gaiani et al. 2007; Mimouni et al. 2009). Mitchell et al. (2015) described a rate-limiting regime mapping approach that describes the particular aspect or regime that limits the overall rehydration in a stirred vessel at a specific condition of agitation and water temperature. This could be due to powder floating on the surface of the water, sediment on the bottom of the vessel, formation of powder clumps or slowly dissolving powder particles remaining dispersed in the water. Fitzpatrick et al. (2019) extended this approach by quantitatively measuring the fraction of powders that float, clump, settle, disperse and dissolve.

The objective of this chapter is to describe a variety of test procedures that can be applied to characterising the rehydration behaviour of food powders, in particular their wettability, solubilisation ability and rehydration behaviour in stirred vessels. Examples are provided where these tests are applied to evaluate and compare the rehydration behaviour of a number of food powders and to investigate how a number of factors influence their rehydration behaviour.

5.2 Powder Wettability

5.2.1 Wetting Time and % Wettability

The powder wetting time is commonly used to characterise powder wettability. Such a test method is described by Fitzpatrick et al. (2016), which is based on the International Dairy Federation method (Anon 1979). This consists of discharging 10 g of powder onto the surface of 150 g of water (at a known temperature) in a glass beaker with a defined internal diameter. The time needed for the powder to immerse below the water surface or be totally wetted is measured. This time is called the wetting time. Powders with wetting times less than 1 min are considered as easy wetting, those with wetting times greater than 20 min are considered as poor wetting, and those with wetting times between 1 and 20 min are considered as intermediate wetting powders.

Some powders may wet very slowly or may not wet fully within 60 min. Fitzpatrick et al. (2016) proposed an extension to this method to assess the wettability of these very poor wetting powders. This involves measuring the amount of powder that does not wet on the water surface after 1, 20 and 60 min. A spoon is used to carefully remove powder from the water surface, which is then dried and weighed to estimate the amount of powder that does not wet. The % wettability is evaluated from Eq. (5.1)

$$\%Wettability = 100 \times \frac{mass\ of\ powder\ wetted}{mass\ of\ initial\ powder} \tag{5.1}$$

This test evaluates the wetting time of a powder if less than 60 min, and also the % wettability at 1, 20 and 60 min for those powders whose wetting time is greater than 1 min.

5.2.2 *Contact Angle*

Wettability is commonly assessed by contact angle measurement using the sessile drop technique (Crowley et al. 2015; Ji et al. 2016). An optical tensiometer (such as Attension Theta, Biolin Scientific Ltd., Espoo, Finland) is used to measure contact angle, whereby a micro droplet of deionised water is deposited onto the surface of the powder. A video or dynamic series of images are taken of the droplet deformation over time, from which the contact angle is evaluated over time. Before the measurement, the powders may first be compacted to produce a tablet. This provides a rigid surface which allows little or no penetration of the water droplet in to the powder. This is appropriate for investigating the influence of powder surface chemistry or composition on powder wettability. Care has to be taken that the compression does not cause expression of any lipids making the surface more hydrophobic. Contact angles greater than 90° are indicative of more poorly wetting powders.

In addition to surface chemistry, the wettability of powders is also influenced by the interstitial pore spaces between the powder particles. This is the reason why an agglomerated powder often has better wettability than the non-agglomerated powder. Compaction of a powder into a tablet will eliminate the effect of the interstitial pores. Consequently, it may not be desirable to produce a tablet. Instead, the powder may be packed into an aluminium pan and a smooth surface is formed using a leveller. The micro-droplet is then deposited onto this surface. In this case, the droplet may change rapidly as it may penetrate into the powder mass and thus the contact angle and its change over time is important. Ji et al. (2016) applied this approach to investigating the influence of agglomeration on the wettability of high protein dairy powders.

5.2.3 *Washburn Method*

The Washburn method is based on capillary rise wetting phenomena, or the ability of a liquid to penetrate up into the interstitial pore spaces between powder particles in a packed powder bed contained in a tube placed above the liquid surface. It usually records the penetration depth of the liquid inside the capillaries as a function of time (Washburn 1921). However, alternative measurements can be applied. Ji et al. (2016) describes an approach whereby a fixed mass of powder is inserted into a cylindrical glass tube with an open base bottom covered by filter paper and gauze to prevent the powder falling out. The tube is set just above the surface of water at a specified temperature. The additional mass of the wetted powder in 10 min, due to water penetration, is used as a quantitative measure of the wettability of the powder.

5.2.4 *Film Strength at Powder/Water Interface*

The wettability of some powders is impeded by the formation of a film or skin at the powder/water interface. Furthermore, this may result in the formation of clumps with slimy films on the outside with dry powder inside. These skins or films can be difficult to disrupt and thus inhibit the wetting of the dry powder. Measurement of the strength of these films could be used as a means for evaluating this aspect of powder wettability, that is, poor wettability due to the formation of these films.

Fitzpatrick et al. (2017) presents a preliminary work on an approach for trying to empirically measure an index of the strength of these films formed at powder/water interfaces. This could be used as an index of a powders susceptibility to poor wetting associated with film formation. This approach applies a force-displacement test to measure the strength of films developed at the powder/water interface. A specified volume of water is filled into a glass vessel. A specified volume of powder is deposited onto the water surface to produce a powder layer of a given thickness. A plunger moves axially through the center of the powder layer at a constant speed, as illustrated in Fig. 5.1a, and force vs displacement is measured. The force-displacement profile is a straight line when there is no powder present. Powders having measurable film strength display a noticeable peak, which is similar to that illustrated in Fig. 5.1b. This is the peak force required for the plunger to penetrate through the film and is used as an index of film strength. Powders displaying negligible film strength display no discernable peak and have a similar linear behavior to that of the plunger moving through water. Measurement is conducted after the powder layer is in contact with water for a specified time.

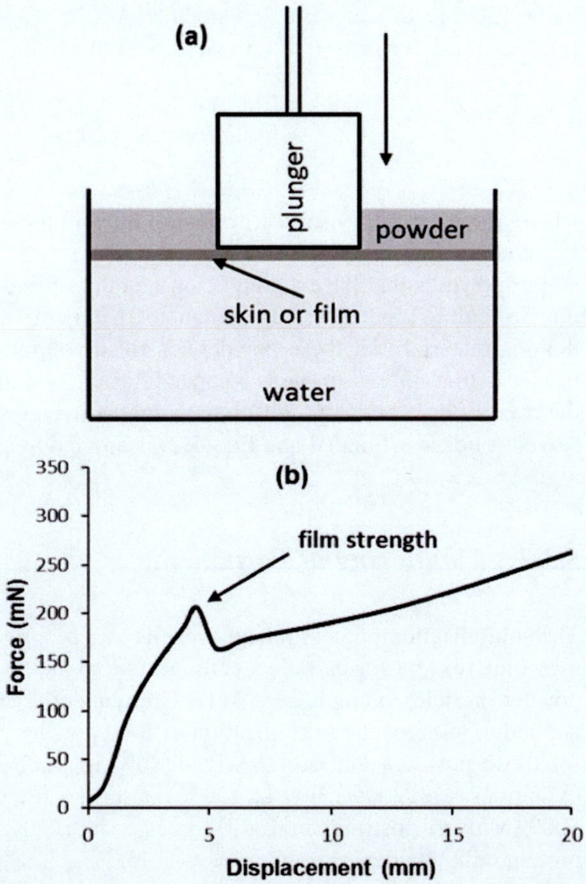

Fig. 5.1 Powder film strength at powder/water interface: (**a**) force-displacement tester; (**b**) film strength of SC powder (20 min, 20 °C)

5.3 Solubilisation Ability

5.3.1 % Solubilisation

Fitzpatrick et al. (2016) describe an approach for measuring the % solubilisation of a powder as a function of time. 150 g of water at a specified temperature is placed in a glass beaker containing an agitator, such as a 6-bladed turbine impeller. To facilitate rapid wetting of the powder, the operating speed is selected so as to create a strong vortex into which 10 g of powder is poured. 5 mL samples are pipetted from a fixed position in the beaker after specific mixing times (1, 5 and 60 min). The samples are deposited into 10 mL tubes which are placed in a centrifuge to settle out any non-solubilised solids. When centrifugation is complete, the supernatant is gently removed and dried. The resultant mass of dried solids is used to measure the concentration of solids in the supernatant, from which the mass of dissolved solids in the 150 mLs of water is calculated (M_{dsf}). The % solubilisation is evaluated using Eq. (5.2), and this is used as an assessment of solubilisation ability.

$$\%solubilisation = 100 \times \left(\frac{M_{dsf}}{M_{dsi}} \right) \tag{5.2}$$

where M_{dsi} mass of dry solids (grams) in the initial mass of powder = $10 \times (1 - W)$; W water content of powder as a mass fraction.

For powders that have a significant amount of particles that float to the surface, it is difficult to pipette the supernatant as the pipette must pass through the layer of floating material. For these powders, a given volume of the mixed contents of the beaker is first filtered through a paper filter with small pore size before measuring the mass of dry solids in the filtrate using the oven test. From this, the mass of dissolved solids is estimated and Eq. (5.2) is applied to estimate the % solubilisation.

5.3.2 *Monitoring of Particle Size*

The solubilisation process usually results in a progressive decrease in particle size over time (except for particle swelling), due to the dissolution of the surface of the powder particles (Fang et al. 2011). Consequently, change in particle size can be applied to monitor the solubilisation of food powder particles over time, especially for those powders that are slowly dissolving, such as casein-dominant powders (Mimouni et al. 2009). Ji et al. (2016) describes the use of a Malvern Mastersizer 3000 (Malvern Instruments Ltd, Worcestershire, UK) to measure the PSD of high protein milk powders every 2 min.

5.3.3 *Centrifugal Sedimentation*

Monitoring of the settling of a suspension in a suitable centrifugal field can be used in the evaluation and comparison of the solubilisation of powders. The basic idea is that powder particles present will settle and cause the suspension to be non-uniform and this can be monitored. The LUMiSizer (L.U.M. GmbH, Berlin, Germany) is an analytical centrifuge that was initially developed and applied to measure the stability of emulsions. It has subsequently been applied to assessing the solubilisation ability of dairy powders (Crowley et al. 2015) by measuring the centrifugal sedimentation of particles present in a measuring cell. This is achieved by measuring the intensity of NIR light transmitted along the length of the measuring cell at discrete times. From the measured light transmission profiles, any sediment height can be evaluated along with an instability index which is an index of the amount of centrifugal sedimentation that takes place.

5.4 Qualitative and Quantitative Assessment of Rehydration Behaviour in a Stirred Vessel

5.4.1 *Rate-Limiting Regime Mapping*

Mitchell et al. (2015) first presented this approach to qualitatively assessing the rehydration behaviour of food powders in a stirred vessel. The experimental set-up consisted of a transparent beaker filled with a given volume of water at a specified temperature. An impeller is located at a fixed position below the water surface. A powder feeder is used to consistently transfer a given mass of powder onto the water surface at a specified mass flowrate. A conductivity probe is used to track dissolution behaviour over time. A camera is used to qualitatively capture the rehydration behaviour of the powder over time. For a given agitation speed, a qualitative assessment is made at discrete times as to what is rate-limiting the rehydration process (if complete dispersion or dissolution is not observed). This assessment uses key words including floating (if powder floats on the surface of the water); and clumps in dispersion (if powder clumps are visible in dispersion); and sedimentation (if powder settles to the bottom of the beaker).

5.4.2 *Quantitative Evaluation of Floating, Clumped, Sedimented, Dispersed and Dissolved Powder*

Fitzpatrick et al. (2019) presented an extension to the assessment presented by Mitchell et al. (2015), by incorporating a quantitative assessment, which provides a more precise numeric characterisation that compliments the qualitative description. This is achieved by measuring the percentage of initial powder added that floats, clumps, settles, disperses and dissolves at given agitation speeds and times.

At the end of each condition of agitation and time in the stirred vessel, powder floating on the water surface is removed and transferred to a pre-weighed glass dish. The remaining suspension is then poured through a sieve into a second beaker, which separates any clumped material remaining in suspension. The size of the sieve aperture can be chosen depending on work undertaken. The sieved powder clumps are transferred to a second pre-weighed glass dish. Any remaining sedimented powder in the first beaker is washed into a third pre-weighed glass dish. The glass dishes are then dried in an oven. After drying, the dry masses of floating powder (M_F), clumped powder in suspension (M_{CS}) and sedimented powder (M_S) are evaluated from the first, second and third glass dishes, respectively. The mass of powder in the dispersion is measured by taking samples from the remaining dispersion and transferring them into pre-weighed glass dishes. The glass dishes are dried for 24 h at 104 °C, from which the mass of powder in dispersion (M_D) is calculated. Finally, a known volume of the dispersion is filtered through a small pore sized filter to estimate the amount of dissolved/finely dispersed material (M_{Dis}) that is in the dispersion.

The relationship between the constituent dry powder masses and the initial dry powder mass added (M_I) is given in Eq. (5.3). The *Error* term represents the difference between M_I and the constituent dry powder masses. If *Error* is positive, this means that some of the initial dry powder added is not accounted for.

$$M_I = M_F + M_{CS} + M_S + M_D + M_{Dis} + Error \tag{5.3}$$

5.5 Examples of the Application of the Characterisation Techniques to Assessing the Rehydration Behaviour of Food Powders

The characterisation techniques described above can be applied to provide a more complete characterisation of the rehydration behaviour of individual powders. They can be applied to investigate how various factors influence rehydration behaviour (such as agglomeration, water temperature and pH), and for comparing the rehydration behaviour of different powders. Below are a number of examples where the rehydration behaviour of different food powders are presented and compared, and where the effects of water temperature and agglomeration on rehydration are investigated.

5.5.1 Comparison of the Rehydration Behaviour of 6 Food Powders

Fitzpatrick et al. (2016) applied some of the techniques outlined above to assessing and comparing the rehydration behaviour of 12 food powders, in terms of their wettability, solubilisation ability and rehydration behaviour in a stirred vessel. In this section, a summary of the results for 6 of the powders (sugar, curry and flour powders, cream milk powder (CMP), sodium caseinate (SC) and milk protein isolate (MPI) powders) is presented. Particle size, apparent density and composition data for these powders are provided in Table 5.1.

5.5.1.1 Wettability

The measured wetting time, % wettability and contact angle of the 6 different food powders are presented in Table 5.2. The sugar and curry powders are easy wetting powders with wetting times less than 1 min, while the SC, MPI and CMP are poor wetting powders with wetting times greater than 60 min. In fact, their % wettability

Table 5.1 Particle size, density and composition of 6 food powders

Powder	Particle size D(50) (μm)	Apparent density (kg L^{-1})	Dry solids composition (%)		
			CHO[a]	Protein	Fat
Sugar	43	1.61	100	0	0
Curry	275	1.34	na[b]	na	na
Flour	85	1.49	87.5	11	1.5
SC	90	1.31	1.5	93	1.5
MPI	50	0.81	0.6	93	1.7
CMP	75	0.93	13	9	75

SC sodium caseinate, *MPI* milk protein isolate, *CMP* cream milk powder
[a]*CHO* carbohydrate
[b]*na* not available

Table 5.2 Wettability of 6 food powders

Powder	Wetting time	% Wettability		Contact angle	
		1 min	60 min	1 s	10 s
Sugar	2 s	100		0	
Curry	25 s	100		28	
Flour	12 min	33	100	38	21
SC	>1 h	20	37	130	129
MPI	>1 h	–	18	150	137
CMP	>1 h	3	5	104	98

at 60 min is low, as presented in Table 5.2. The flour powder could be considered as an intermediate wetting powder, having a wetting time of 12 min. The contact angle data relate closely to the wetting time and % wettability data. The sugar, curry and flour powders had contact angles less than 90°, while all the poor wetting powders had contact angles greater than 90°. Furthermore, after 10 s, there was no contact angle for the easy wetting powders, and poor wetting powders had contact angles in excess of 90°. Surface hydrophobicity is determined by surface composition and has a major influence on contact angle, consequently composition is having a major influence on the wettability of the powders.

The sugar (sucrose) wetted almost instantly, which is to be expected due to sucrose being highly hydrophilic and the higher density of the sugar crystals. The curry powder had good wettability with a wetting time of 25 s, which is also influenced by its relatively large particle size (Table 5.1). The flour powder had a wetting time of 12 min, which is possibly due to it having higher surface hydrophobicity than the sugar and curry, as evidenced by the contact angle data.

The CMP essentially did not wet, which is due to the hydrophobic nature of its major constituent, fat, and its apparent density being lower than that of water. SC had poor wettability, which is most likely due to its composition consisting mainly of casein proteins, and the hydrophobic nature of the particle surfaces, as evidenced by its high contact angles. MPI powder displayed poor wettability, which is also due to its composition containing mainly casein proteins, resulting in a hydrophobic surface as evidenced by its very high contact angle. Furthermore, its apparent density was less than that of water and its particle size was small at 50 microns. This would suggest that the powder should float in the wettability test and not sink, thus this may partially explain the poor wettability.

5.5.1.2 Solubilisation Ability

The % solubilisation ability of the powders is presented in Table 5.3. The powders display differing solubilisation behaviours, which can be broadly classified as:

Easy to Solubilise The sugar solubilised very quickly as expected. The SC powder can also be regarded as being easy to solubilise as 80% of it readily solubilised, however the remaining 20% more slowly solubilised over time.

Contains a High Insoluble Fraction These are powders that have an inherent amount of insoluble material which readily settled out during centrifugation. The soluble content solubilised rapidly, as the % solubilisation attained a near constant value within 1 min. The powders included the curry, flour and CMP. The curry powder had about 32% solubles which rapidly solubilised, however the remaining 68% is essentially insoluble material. The flour is made up mainly of insoluble fine particulates of milled materials of plant origin. The CMP powder has 73% fat which is inherently insoluble in water leading to low solubilisation ability.

Slow to Solubilise These are powders that solubilise slowly over time. They included the MPI in particular, which slowly solubilised over time with 44% solubilised after 1 h but it does eventually fully solubilise over time. Furthermore, the type and intensity of agitation and temperature can quicken its rate of solubilisation.

5.5.1.3 Qualitative Rate-Limiting Regime Mapping in a Stirred Vessel

The rate-limiting regime mapping for the sugar, curry and flour powders is presented in Table 5.4. Sedimentation was the rate limiting regime for these powders at low agitation (200 rpm and less). The curry powder was either solubilised or well dispersed at speeds of 300 rpm and higher. The sugar encountered the formation of some clumps in dispersion at higher speeds of 300–500 rpm, which was most likely due to the much smaller particle size. This made water penetration into the powder

Table 5.3 % Solubilisation of 6 food powders

	% Solubilisation					
Powder	Sugar	Curry	Flour	SC	MPI	CMP
1 min	97	31	4	80	23	15
60 min	100	32	589	89	44	20

more difficult due to the small interstitial voids between powder particles. The flour powder displayed a more complex rate-limiting regime map. It displayed both floating and sedimentation up to 300 rpm. Floating was more dominant at lower speed, but even at 100 rpm, some lumps of powder sedimented to the bottom and this increased with higher speed. At 400 and 500 rpm, floating disappeared and the rate-limiting regime was due to clumps either sedimenting to the bottom or becoming dispersed in the water. At 600 and 700 rpm, sedimentation ceased and the rate-limitation was due to smaller clumps in suspension. The powder only dispersed well at the highest speed of 750 rpm.

The rate-limiting regime mapping of the poor wetting SC, MPI and CMP powders is presented in Table 5.5, where floating was the dominant at lower speeds. Very little sinking and dispersion was displayed by CMP up to 300 rpm. This is expected due to its inherently hydrophobic nature and having an apparent density less than the density of water. Development of a stronger vortex at 400 rpm forced more powder and fine clumps into dispersion but there were still a lot of clumps of powder circulating in the vortex. Higher speeds reduced the presence of floating clumps although some were still present at 750 rpm along with fine clumps in dispersion. Floating was the dominant rate-limiting regime for both the MPI and SC powders, most likely due to high hydrophobicity caused by high casein protein content (Table 5.1), and the low apparent density of the MPI powder. The MPI powder behaved somewhat similar to the CMP in the sense that increased dispersion occurred at higher speeds and complete dispersion was only achieved at the highest speed of 750 rpm.

For the SC powder at 500 rpm and higher, powder was submerged into the water as clumps by the stronger vortex, and clumping became rate-limiting, even at the highest speed of 750 rpm where some powder clumps in dispersion were still evident. The surface of the clumps had a strong skin or film with dry powder inside that greatly impeded waters ability to penetrate into the bulk. The presence of strong film formation made it difficult for the clumps to be disrupted during agitation. Quantitative assessment was performed on the rehydration of the SC powder. This showed that although the clumps had a major visual impact, they accounted for only a small percentage of the powder, typically being less than 5%, and that over 90% of the powder was well dispersed.

Table 5.4 Rate-limiting regime mapping for easy to intermediate wetting powders (sugar, curry and flour powders)

Speed (rpm)	Sugar	Curry	Flour
100	Sedimentation	Sedimentation	Floating + sedimentation
200			
300	Clumps in dispersion	Well-dispersed	
400			Clumping + sedimentation
500			
600	Well-dispersed		Small clumps in dispersion
700			
750			Well-dispersed

Table 5.5 Rate-limiting regime mapping for poor wetting powders (CMP, SC and MPI powders)

Speed (rpm)	CMP	SC	MPI
100	Floating	Floating (*decreased with agitation*)	Floating (*decreased with agitation*)
200			
300			
400	Floating (*decreased with agitation*) fine clumps in dispersion		
500		Big clumps in dispersion	
600			
700			
750			Well-dispersed

5.5.2 *Effect of Temperature on the Rehydration Behaviour of SC and CMP*

5.5.2.1 Qualitative and Quantitative Assessment in a Stirred Vessel

Various factors may influence the rehydration behaviour of a powder. Rate-limiting regime mapping can be applied to investigate their effects. The mapping approach and quantitative assessment were applied to highlight this by investigating the effect of temperature on the rehydration behaviour of SC and CMP in a stirred vessel at 20 °C and 70 °C (Fitzpatrick et al. 2019). It was visually observed that lower temperature resulted in better rehydration behaviour for the SC powder. A quantitative assessment comparison at 650 rpm is presented in Table 5.6, which shows that the powder was well dispersed, except for a small amount of floating powder and a small amount of clumping. At 70 °C, floating became dominant with 81% floating at 20 s, while the powder only became well dispersed at 10 min. There was also no clump formation in suspension at 70 °C. For CMP, the higher temperature improved the rehydration behaviour. This can be seen in the quantitative assessment comparison (Table 5.6), where there was less floating at 70 °C and correspondingly greater dispersion at each of the agitation times. Complete dispersion was achieved at 10 min at 70 °C, while less than 50% was dispersed at 20 °C for the same agitation time.

5.5.2.2 Film Strength at Powder/Water Interface

Poor wetting powders like SC and CMP tend to display floating and clumping behaviours. Hydrophobic surfaces and slimy skin or film formation can cause both floating and clumping. For the SC powder, the presence of clumps in dispersion was noticeable, and the clumps had strong surface films making them difficult to disrupt. Consequently, this section assesses the film strength developed by the SC powder and CMP, and how it is influenced by temperature.

Fitzpatrick et al. (2017) empirically measured film strength at the powder/water interface for a number of food powders, and the force displacement curves for SC

Table 5.6 Effect of temperature on the quantitative rate-limiting regime mapping at 650 rpm for SC and CMP

	SC				CMP			
	20 °C		70 °C		20 °C		70 °C	
Dry mass (%)	20 s	10 min	20 s	10 min	20 s	10 min	20 s	10 min
M_F	9.4	0.0	81.0	0.0	93.8	67.2	63.3	0.0
M_{CS}	3.4	2.1	0.0	0.0	0.0	0.0	0.0	0.0
M_D	88.0	96.6	15.3	99.6	5.0	34.0	36.7	98.9
Error	−0.7	1.3	3.7	0.4	1.2	−1.3	0.01	1.1

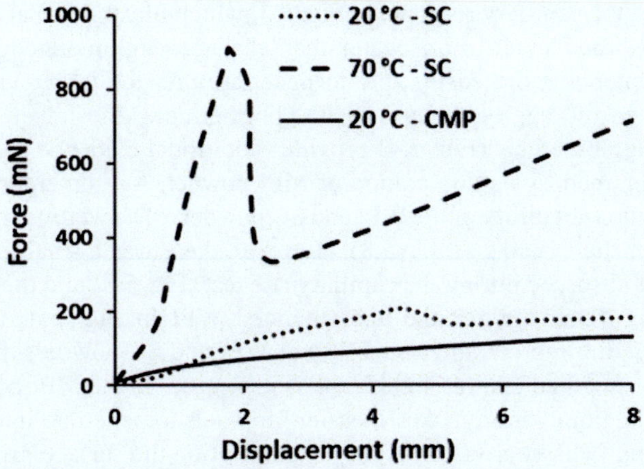

Fig. 5.2 Effect of temperature on film strength of SC powder and CMP at 20 min contact time

and CMP are presented in Fig. 5.2. The film strength of the CMP was negligible at contact times of 1 and 20 min, consequently, strong film formation did not majorly impede the wetting of CMP. On the other hand, the SC powder displayed measurable film strengths. At 20 °C, the film strength was about 200 mN at 20 min contact. Increasing the temperature to 70 °C had a major impact on film strength whereby it increased to 900 mN. This concurs with that visually observed in the rate-limiting regime mapping, and suggests that the SC powder has a propensity to develop very strong films at higher temperatures.

5.5.3 *Effect of Agglomeration on the Rehydration Behaviour of MPI, SC and WPI Powders*

Ji et al. (2016) performed a study that investigated the effect of fluid-bed agglomeration on the rehydration ability of high protein dairy powders including MPI, SC and whey protein isolate (WPI) powders. This section summarises the key results using the characterisation methods outlined earlier. The median particle sizes of the non-agglomerated and agglomerated powders are presented in Table 5.7

5.5.3.1 Effect of Agglomeration on Wettability

SC, WPI and MPI powders are all poor wetting powders, as evidenced by their long wetting times (Table 5.8), poor capillary rise values (Fig. 5.3) and their large contact angles (Fig. 5.4). Fluid-bed agglomeration greatly improved the wettability of the MPI powder, as shown by a major reduction in the wetting time (Table 5.8). This is re-inforced by the capillary rise data in Fig. 5.3, which illustrates that the agglomerated MPI absorbed much more water than the non-agglomerated powder. This shows that water is more easily able to penetrate into the larger interstitial void spaces between the agglomerate particles. Furthermore, the lower contact angle data for the agglomerates (Fig. 5.4) provides additional evidence to the improved wettability obtained by agglomeration of MPI powder. Agglomeration had less of an impact on the wettability of the WPI and SC powders. The wetting times remained long at greater than 1200 s (Table 5.8). However, there was a small increase in the mass of water absorbed during the capillary rise test (Fig. 5.3) and the contact angle was lower, as illustrated for the agglomerated WPI in Fig. 5.4. Consequently, agglomeration did improve the wettability of WPI and SC powders, but the impact was relatively small and not as significant as its impact on the MPI powder.

Although agglomeration makes it easier for water to penetrate into a bulk powder, the wetting behaviour is still highly dependent on the surface properties of the powder. This includes hydrophobicity and also the propensity of a powder to form slimy skins or films on the powder/water interface. In the previous section, it was shown that SC forms such strong films. WPI powder also forms strong films (Ji et al. 2017), and this propensity to form strong films is possibly the reason why water penetration into the bulk was inhibited, resulting in agglomeration being less effective in improving the wettability of the WPI and SC powders.

5.5.3.2 Effect of Agglomeration on Solubilisation Ability

Median particle sizes of the non-agglomerated and agglomerated powder particles during solubilisation are presented in Fig. 5.5. The non-agglomerated WPI and SC powders solubilised rapidly into water [data for the WPI is not shown because the

Table 5.7 Median particle sizes D(50) of the non-agglomerated (NA) and agglomerated (A) high protein milk powders

D(50) (μm)			
	MPI	WPI	SC
NA	49	55	85
A	188	180	208

Table 5.8 Wetting time of non-agglomerated (NA) and agglomerated (A) high protein milk powders

		MPI	WPI	SC
Wetting time (s)	NA	>1200	>1200	>1200
	A	480	>1200	>1200

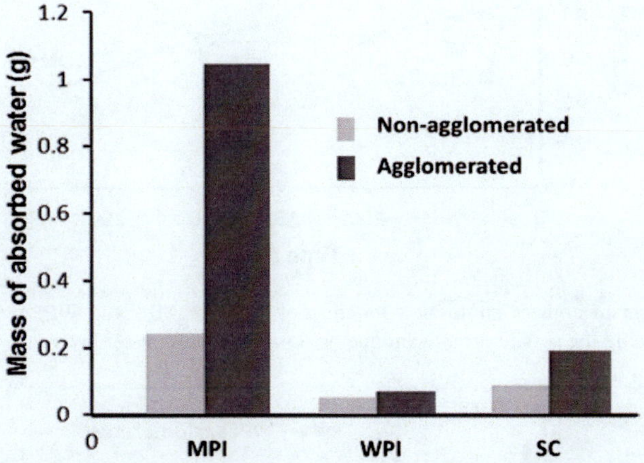

Fig. 5.3 Mass of water sorbed during capillary rise wetting after 10 min by MPI, WPI and SC powders and their agglomerates

powder had solubilised before the first size measurement was taken at 2 min]. As the non-agglomerated WPI and SC solubilised rapidly, the effect of agglomeration was not significant. The MPI powder is a slowly dissolving powder and agglomeration had little effect, as illustrated in Fig. 5.5. In fact, agglomeration caused the solubilisation to be slightly slower. This is because additional time is required to disaggregate the agglomerates into the primary particles before the primary particles can solubilise. The results of the centrifugal sedimentation testing carried out in the LUMiSizer are presented in Table 5.9. These results confirm the easy solubilisation of both the agglomerated and non-agglomerated WPI and SC powders, as evidenced by the measurement of no sediment, and the slow to solubilise behaviour of the MPI powder. Overall, the potential benefit of agglomeration is in improving the wettability of the powders.

5.6 Conclusions

This chapter outlines a variety of techniques that can be applied in combination for providing a more complete characterisation of the rehydration behaviour of food powders. The different techniques can complement each other, whereby some techniques can reinforce each other's findings and other techniques can provide information on different aspects of the rehydration behaviour of a powder. The different

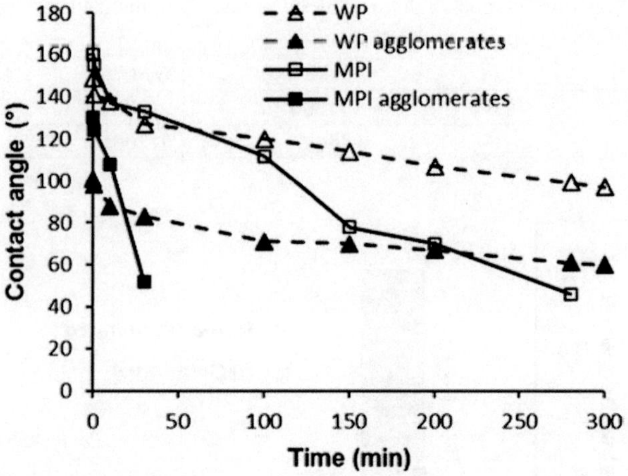

Fig. 5.4 Change in contact angle as a function of time for MPI and WPI powders and their agglomerates using the sessile drop technique on non-tabletted powder layer

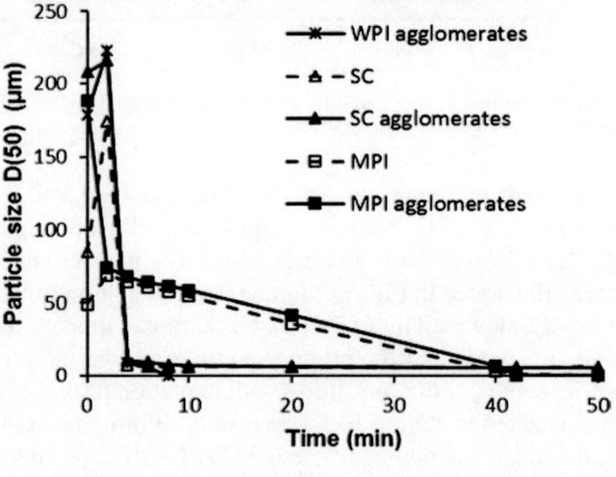

Fig. 5.5 Particle size D(50) measurements of dispersed particles of MPI, SC and WPI powders and their agglomerates (There was no particle size measured for non-agglomerated WPI as it quickly dissolved into water)

Table 5.9 Sediment height after 168 g centrifugation for 10 min for non-agglomerated (NA) and agglomerated (A) high protein milk powders

		MPI	WPI	SC
Sediment height (mm)	NA	2.45	0	0
	A	1.95	0	0

techniques can be applied for investigating how the manipulation of various factors may be applied to improve the rehydration behaviour a particular powder, and for comparing the rehydration behaviour of different powders.

References

Anon. (1979). Proceedings of the IDF seminar on dairy effluents (1976). *International Journal of Dairy Technology, 32*(2), 113–113. https://doi.org/10.1111/j.1471-0307.1979.tb01909.x.

Barbosa-Canovas, G. V., Ortega-Rivas, E., Juliano, P., & Yan, H. (2005). *Food powders: Physical properties, processing, and functionality*. New York: Kluwer Academic.

Crowley, S. V., Desautel, B., Gazi, I., Kelly, A. L., Huppertz, T., & O'Mahony, J. A (2015). Rehydration characteristics of milk protein concentrate powders. *Journal of Food Engineering, 149*, 105–113. https://doi.org/10.1016/j.jfoodeng.2014.09.033.

Dupas, J., Verneuil, E., Ramaioli, M., Forny, L., Talini, L., & Lequeux, F. (2013). Dynamic wetting on a thin film of soluble polymer: Effects of nonlinearities in the sorption isotherm. *Langmuir, 29*(40), 12572–12578. https://doi.org/10.1021/la402157d.

Fang, Y., Selomulya, C., Ainsworth, S., Palmer, M., & Chen, X. D. (2011). On quantifying the dissolution behaviour of milk protein concentrate. *Food Hydrocolloids, 25*(3), 503–510.

Fitzpatrick, J. J., van Lauwe, A., Coursol, M., O'Brien, A., Fitzpatrick, K. L., Ji, J., & Miao, S. (2016). Investigation of the rehydration behaviour of food powders by comparing the behaviour of twelve powders with different properties. *Powder Technology, 297*, 340–348. Retrieved from https://linkinghub.elsevier.com/retrieve/pii/S0032591016302042.

Fitzpatrick, J. J., Salmon, J., Ji, J., & Miao, S. (2017). Characterisation of the wetting behaviour of poor wetting food powders and the influence of temperature and film formation. *Kona Powder and Particle Journal, 34*, 282–289. https://doi.org/10.14356/kona.2017019.

Fitzpatrick, J. J., Bremenkamp, I., Wu, S., & Miao, S. (2019). Quantitative assessment of the rehydration behaviour of three dairy powders in a stirred vessel. *Powder Technology, 346*, 17–22. https://doi.org/10.1016/j.powtec.2019.01.087.

Forny, L., Marabi, A., & Palzer, S. (2011). Wetting, disintegration and dissolution of agglomerated water soluble powders. *Powder Technology, 206*(1–2), 72–78.

Freudig, B., Hogekamp, S., & Schubert, H. (1999). Dispersion of powders in liquids in a stirred vessel. *Chemical Engineering and Processing Process Intensification, 38*(4–6), 525–532. https://doi.org/10.1016/s0255-2701(99)00049-5.

Gaiani, C., Schuck, P., Scher, J., Desobry, S., & Banon, S. (2007). Dairy powder rehydration: Influence of protein state, incorporation mode, and agglomeration. *Journal of Dairy Science, 90*(2), 570–581. https://doi.org/10.3168/jds.s0022-0302(07)71540-0.

Gaiani, C., Scher, J., Schuck, P., Desobry, S., & Banon, S. (2009). Use of a turbidity sensor to determine dairy powder rehydration properties. *Powder Technology, 190*(1–2), 2–5. https://doi.org/10.1016/j.powtec.2008.04.042.

Goalard, C., Samimi, A., Galet, L., Dodds, J. A., & Ghadiri, M. (2006). Characterization of the dispersion behavior of powders in liquids. *Particle and Particle Systems Characterization, 23*(2), 154–158. https://doi.org/10.1002/ppsc.200601024.

Ji, J., Cronin, K., Fitzpatrick, J., & Miao, S. (2017). Enhanced wetting behaviours of whey protein isolate powder: The different effects of lecithin addition by fluidised bed agglomeration and coating processes. *Food Hydrocolloids, 71*, 94–101.

Ji, J., Fitzpatrick, J., Cronin, K., Maguire, P., Zhang, H., & Miao, S. (2016). Rehydration behaviours of high protein dairy powders: The influence of agglomeration on wettability, dispersibility and solubility. *Food Hydrocolloids, 58*, 194–203. Retrieved from https://linkinghub.elsevier.com/retrieve/pii/S0268005X16300704.

Mimouni, A., Deeth, H. C., Whittaker, A. K., Gidley, M. J., & Bhandari, B. R. (2009). Rehydration process of milk protein concentrate powder monitored by static light scattering. *Food Hydrocolloids, 23*(7), 1958–1965. https://doi.org/10.1016/j.foodhyd.2009.01.010.

Mitchell, W. R., Forny, L., Althaus, T. O., Niederreiter, G., Palzer, S., Hounslow, M. J., & Salman, A. D. (2015). Mapping the rate-limiting regimes of food powder reconstitution in a standard mixing vessel. *Powder Technology, 270*, 520–527. https://doi.org/10.1016/j.powtec.2014.08.014.

Schubert, H. (1993). Instantization of powdered food products. *International Chemical Engineering, 33*(1), 28–45.

Washburn, E. W. (1921). The dynamics of capillary flow. *Physics Review, 17*(3), 273.

Chapter 6
Anticaking Additives for Food Powders

Emine Yapıcı, Burcu Karakuzu-İkizler, and Sevil Yücel

6.1 Caking Formation in Food Powders

Most powder materials tend to form lumps for various reasons and cause many unwanted problems. It is essential that understand caking phenomena and approaches to prevent the formation of caking to overcome all problems related to lumping (Irani et al. 1959). Moisture content, temperature, pressure, impurities and storage time are important parameters in caking experiments. Studies were performed to observe the effects of different combinations of parameters on caking behavior (Irani et al. 1959).

Caking is an undesirable phenomenon that free-flowing powders come together and turn into a solid cake during storage (Hansen et al. 1998). The lumping of low-moisture powder materials by agglomeration and getting sticky, hence the material loses its functionality and also decreases its quality and yield. Powders also lose their free flowability, flavor, and crispness because of caking. This is an economically important problem in the related industries where powder materials are used (Aguilera et al. 1995).

During the typical caking process, bridging, agglomeration, compaction and liquefaction steps take place, respectively. Lumps can form in different amounts, various sizes, and varying hardness in each stage of caking. Follow-up of morphological changes occurring in the particle diameter and the porosity of the particles with the formation of a bridge between particles provides a quantitative measurement of caking (Aguilera et al. 1995).

Bridging, the first step of caking, occurs as a result of the surface deformation and adhesion at the contact points between particles, without a measurable reduction in system porosity. Formed bridges can be broken even under an impact at this stage of caking. In the agglomeration phase, despite the irreversible consolidation of

E. Yapıcı (✉) · B. Karakuzu-İkizler · S. Yücel
Department of Bioengineering, Faculty of Chemistry and Metallurgy, Yildiz Technical University, Istanbul, Turkey

E. Ermiş (ed.), *Food Powders Properties and Characterization*, Food Engineering Series, https://doi.org/10.1007/978-3-030-48908-3_6

the bridges, the high porosity of the particle system is maintained and particle lumps with structural integrity are formed. In the compaction step, interparticle bridges become thicker, the interparticle gaps are reduced, and a significant loss of system integrity occurs as a result of the particle aggregates deformation under pressure. The final step of caking, liquefaction, interparticle bridges completely disappears as a result of high moisture content and dense flow (Aguilera et al. 1995).

The situations to be controlled and methods used to prevent caking can be listed as follows.

- decreasing the fine particle content of the powder;
- decreasing the moisture content of the powder;
- identifying the major caking component and changing if an alternative is available;
- improving storage condition with reducing temperature and humidity;
- reducing the time consolidation load;
- usage of anticaking agents (Zafar et al. 2017)

The flowability and flow behavior of the powders directly related to the caking process. Caking formation and the flowability loss of powder materials during storage and transportation cause reduced quality (Ganesan et al. 2008).

Various precautions should be taken to reduce caking during storage of food powders and similar materials. Several factors affect flowability and accelerate the caking process of powders including moisture content, temperature, pressure, fat amount, particle size, and anticaking agents (Juliano and Barbosa-Canovas 2010; Ermiş et al. 2018). These factors must be managed to reduce the tendency of a food powder caking.

Moisture control is a major factor to inhibit microbial growth problems in food powders. Food powders are mostly hygroscopic and tend to adsorb moisture in suitable humidity conditions. When water sorption increases, adhesive and cohesive forces resulting in liquefication between particles also increase.

Temperature is another essential factor that affects the flowability properties of food powder. The temperature of amorphous or semi-crystalline food systems determines whether the mobility of food molecules is glassy or viscous. The viscosity of food systems is a function of temperature. If the temperature of amorphous foods is higher than the glass transition temperature (Tg), they become liquid-like, rubbery associated with stickiness and agglomeration. This means that if the product temperature is lower than the glass transition temperature, events such as caking and sticking will not occur (Taylor et al. 1991).

6.2 Chemical Substances in Functional Class: Anticaking Agents

Flow conditioners or anticaking agents are the additives that provide a steady flow of any powder mass and facilitate the flow in the storage container (Irani et al. 1959). Anticaking agents are frequently used as food additives that can help a powder in maintaining a steady flow and/or increase its flow rate. Anticaking agents improve the flowability of the powders by reducing the stickiness and compressibility of the intergranular forces. They are also known as flow conditioners, free-flowing agents, antistick agents, lubricants, glidants, drying agents, dusting powder, release agents because caking is also related to flowability and stickiness properties during storage (Branen et al. 2002). Anticaking additives become important especially in the pharmaceutical, chemical, food (Chang et al. 2018; Hollenbach et al. 1982), fertilizer (Martinez and Rocafull 2015) and feed (Rychen et al. 2017) industries due to the content of powder materials.

Anticaking additives are generally very fine powders that measured in micron and fewer particle sizes (approximately 40–100 μm) (Juliano and Barbosa-Canovas 2010). They are also chemically inert substances. Anticaking additives can be synthetic, but most of them are natural or nature identical. Silicates, polysaccharides, phosphates, stearates, and iron salts are the common types of anticaking agents. Most of them are insoluble or partially soluble in water and ethanol. One of the common features of anticaking agents is having a high surface area. In this way, these agents can adsorb significant amounts of water as an advantage of having large surface areas. It should be noted that anticaking agents must finer particle size than host powder. Caking formation of host powder can be prevented better when using the finer particle size of the anticaking agent (Irani et al. 1959; Ganesan et al. 2008).

The working efficacy of additives changes in a variety of powder materials and different environmental conditions. Therefore, the optimum working conditions of additives can be determined by experimental studies.

Effective moisture content control and storage at low temperatures as possible are important factors to minimize the agglomeration of powders. However, in many cases, additives are added to hygroscopic food powders to increase flowability and/or prevent cake. Flow conditioners or anticaking agents work in slightly several mechanisms to overcome the caking problems.

One of the most important abilities of these materials is to compete with the host powder for the existing moisture in the environment. Most of the agents protect foods by absorbing large amounts of water vapor due to their porous structure.

They become physical surface obstacles between food powders, increase the distance and reduce friction between particles. They interfere with liquid bridging mechanisms. They inhibit crystal growth by enabling the reduction or neutralization of surface molecular attractive electrostatic forces with opposite charges (Lück et al. 2012).

Increasing the glass transition temperature (Tg) of the amorphous phase prevents the liquid-like powders associated with stickiness and agglomeration. Anticaking agents play an important role in increasing the stability of host powders by increasing the Tg of the powder (Chang et al. 2018). Anticaking agents also prevent caking on the surface of the hygroscopic particles creating a moisture-protective barrier without entering the amorphous phase of the powder (Aguilera et al. 1995).

Anticaking agents are widely used in many foods in powder and granular form. These can be listed as vegetable, beverage, and fruit powders, powdered egg, powdered soups, yeast powder, confectionery products, vending machine powders (milk, coffee, cream powders), grated cheese, powdered flavors, salt and spices, powdered sauces, topping powders, baking powder and cake mixes, icing sugar and powdered chocolate (Lück et al. 2012).

Anticaking agents are frequently added in powder systems to delay or prevent caking, but there is insufficient knowledge about their effect on the chemical and physical stability of powders. There is no specific analysis method for evaluating the performance of anticaking agents. The main reason for this problem is that several independent variables affect the tendency of powders caking (humidity, temperature, host powder properties, etc.) (Lipasek et al. 2011; Hollenbach et al. 1982).

6.3 Commonly Used Anticaking Agents

6.3.1 Aluminum Silicates

According to E numbers, aluminum silicates can list as E554 sodium aluminum silicate, E555 potassium aluminum silicate, E556 calcium aluminum silicate, E559 aluminum silicate (kaolin). Aluminum silicates are commonly obtained by precipitation of soluble aluminum salts and suitable metal. They are often described as white, amorphous, powder materials. Aluminum silicates are fine powders used with food as free-flowing agents in the food industry. It is used in beverage powders and sweet powders. Aluminum silicates are relatively inexpensive compared to other anticaking materials that control flow behavior. It is often preferred due to its affordable price and its performance in improving powder flow (Emerton et al. 2008).

United Nations Joint Food and Agriculture Organization/World Health Organization Food Additives Expert Committee (JECFA), 1 mg/kg to receive acceptable daily intake (acceptable daily intake-ADI) in 2006, temporarily tolerable weekly intake (Provisional tolerable weekly intake-PTWI) determined 7 mg/kg body weight. These rates apply to all Al compounds in food, including additives (Pandey and Upadhyay 2012).

Calcium aluminum silicate is defined as fine, white, free-flowing powder. It should contain 44–50% silicon dioxide (SiO_2), 3–5% aluminum oxide (Al_2O_3), 32–38% calcium oxide (CaO), 0.5–4% sodium oxide (Na_2O) (JECFA 2019).

Sodium aluminum silicate, According to FAO JECFA, is defined as an amorphous hydrated sodium aluminum silicates with varying proportions of Na_2O, Al_2O_3, and SiO_2. It is obtained by reacting aluminum sulphate and sodium silicate followed by precipitation method. Sodium aluminum silicate is described as a fine, white, free-flowing powder. Silicon dioxide (SiO_2) content of the sodium aluminum silicate should not less than 66% and not more than 88%. Also, Aluminium oxide (Al_2O_3) content of it should not less than 5%, not more than 15%. And Sodium oxide (Na_2O) content should not less than 5% and not more than 8.5% (JECFA 2019).

The use of sodium aluminum silicate is more common than potassium aluminum silicate. Potassium aluminum silicate is often used as a carrier. Besides, potassium aluminum silicate is used where appropriate to reduce the sodium content of the foodstuff (Emerton et al. 2008).

6.3.2 *Bentonite E558*

Chemical composition of Bentonite samples mainly consists of silicon dioxide (SiO_2), magnesium oxide (MgO), aluminum oxide (Al_2O_3) and sodium oxide (Na_2O). The percentage values (%) of the chemical elements diverge according to the type of bentonite samples. There are not determined elemental amounts according to FAO (Food and Agriculture Organization). The results of various studies are examined and it is seen that bentonite can contain 51.5–72.4% silicon dioxide (SiO_2), 2.6–26.1% magnesium oxide (MgO), 4.1–23.8% aluminum oxide (Al_2O_3) and 1.2–3.1% sodium oxide (Na_2O) (EFSA 2012).

There are no restrictions on the use of bentonites as anticaking agents in foods. Bentonites are safe and specially used in animal feed as a feed additive (Rychen et al. 2017). According to the results of the Panel on Additives and Products or Substances used in Animal Feed (FEEDAP), it was concluded that bentonite at 20,000 mg/kg of complete feed did not produce any safety consequences for all animals and consumers.

6.3.3 *Calcium Carbonate E170 (i) and Calcium Hydrogen Carbonate E170 (ii)*

Calcium carbonate is a permitted food additive used to improve many properties of foods. It can be added to foods for several purposes such as acidity regulator, food coloring, anticaking agent, etc. It is also known as carbonic acid calcium salt, calcite, and chalk. Calcium carbonate ($CaCO_3$) is an odorless, white inorganic salt and its molecular weight is 100.1 g/mol. Calcium carbonate has six solid forms including micro-crystalline (anhydrous crystalline; calcite, aragonite, vaterite and hydrated crystalline; crystalline monohydrocalcite and ikaite) or amorphous powder (Opinion 2011). The particle diameter of amorphous spherical calcium carbonate is in the range typically 40–120 nm, crystals forms of calcium carbonate particles have generally in range 1–10 μm diameter (Meiron et al. 2010). Nanoparticle size calcium carbonate is not suitable for use as a food additive. The average particle size (d50) of food-grade calcium carbonate can be about 5 μm and upper limit (d98) of 65 μm (Opinion 2011).

Calcium hydrogen carbonate is also known as calcium bicarbonate. Its chemical formula is Ca $(HCO_3)_2$ and molecular weight is 162.1 g/mol. Calcium hydrogen carbonate is a white crystalline powder that is soluble in water, 16.6 g/100 mL (20 °C). Calcium bicarbonate can be produced by a reaction between calcium carbonate and carbonic acid. The reverse process, when heated, calcium bicarbonate decomposes into calcium carbonate, carbon dioxide, and water. It is used as a color stabilizer, anticaking food additive in foods. There is no detail information about calcium bicarbonate (NCBI 2020a, b).

6.3.4 Calcium Silicate E552

Calcium silicate is a hydrous or anhydrous inorganic material used as an anticaking agent in the food industry. It is a water-insoluble, white-colored and very fine powder. Calcium silicate has also a low bulk density and a high water absorption capacity. It is prepared by various reactions between siliceous material and calcium compounds. It can be obtained from naturally occurring limestone and diatomaceous earth or produced synthetically from silicon dioxide and calcium oxide with various ratios. Calcium silicate is an inorganic substance that is a hydrous or anhydrous substance with varying proportions of calcium as calcium oxide, and silicon as silicon dioxide. According to FAO assay that calculated on the ignited basis, calcium silicate should contain from 50% to 95% silicon dioxide (SiO_2), from 3 to 35% calcium oxide (CaO) (JECFA 2019). It is proper to use in foods not exceeding 2% by weight and not exceeding 5% by weight in baking powder (NCBI 2020c).

6.3.5 Ferric Ammonium Citrate E381

Ferric Ammonium Citrate is also known as iron ammonium citrate, ammonium ferric citrate, ammonium iron citrate, ammonium iron (III) citrate. It is a complex salt with an indeterminate structure consisting of iron, ammonia and citric acid. It is referred to as brown and green salt according to the iron content in different amounts. It is stated that it may contain 16.5–22.5% iron for brown salt and 14.5–16.0% iron (Fe) for green salt. Green salt is especially used as an anticaking agent (JECFA 2019). The usage level of iron ammonium citrate in the host powder salt should not exceed 0.0025% by weight (WHO 2006).

6.3.6 Isomalt E953

Isomalt is a white powder, odorless, crystalline and weak hygroscopic substance. In addition to being frequently used as a synthetic sweetener in foods, isomalt is also used as a bulking agent, preventive agent and glazing agent. Chemical names and formulas of Isomalt types are 6-O-alpha-D-Glucopyranosyl-D-sorbitol; $C_{12}H_{24}O_{11}$,

1-O-alpha-D-Glucopyranosyl-D-mannitol dihydrate; $C_{12}H_{24}O_{11}{\cdot}2H_2O$. Isomalt can also be used as an anti-caking agent in ready-to-eat cereal products, sugar free confectionery, frozen foods such as ice cream, fish and meat products (McNutt and Sentko 2003).

6.3.7 Magnesium Hydroxide Carbonate E554 (ii)

Magnesium hydroxide carbonate is an odorless, light, friable or a bulky-white powder. It is also referred to as magnesium subcarbonate (light or heavy), hydrated basic magnesium carbonate, magnesium carbonate hydroxide. The magnesium oxide (MgO) content should not be less than 40% and more than 45% (JECFA 2019). In addition to being used as an anticaking agent in foods, it is also used as a drying agent, color retention agent, and carrier.

6.3.8 Magnesium Oxide E530

Magnesium oxide (MgO) is used as an anticaking agent in foods. It is named according to density as light magnesium oxide (0.1–0.12 g cm^{-3}) and heavy magnesium oxide (0.25–0.5 g cm^{-3}). After burning at approximately 800 °C, magnesium oxide content should not be less than 96.0%. It is insoluble in ethanol and water (JECFA 2019).

6.3.9 Magnesium Silicate E553a

Food additive magnesium silicate E553 a according to European Commission Regulation No 231/2012; It is defined as a synthetic compound with a mole ratio of magnesium oxide to silicon dioxide approximately 2:5. According to the same commission, it is defined as a very fine, white, odorless powder material. MgO content is more than 15% and SiO_2 content is more than 67%.

The production of high quality and homogeneously dispersed silicate, the reaction of sodium metasilicate solution is carried out with solutions of suitable salts, and it is obtained by the precipitation of precipitates (Younes et al. 2018c).

Magnesium silicate is used in confectionery products as an anti-stick and anti-caking agent (a component of molding powder or anti-gloss paste). Magnesium silicates serve also as a carrier and preventive substances in vitamin and mineral premix preparations of animal feeds. It is possible to use white pigment instead of titanium dioxide as it can easily provide white color (Rashid et al. 2011).

6.3.10 Silicon Dioxide E551

According to the Joint FAO/WHO Expert Committee on Food Additives (JECFA), silicon dioxide is defined as an amorphous substance called fumed silica or hydrated silica according to the production method. Chemically formulated as $(SiO_2)_x$ and the molecular weight is 60.08 g/mol.

Two different manufacturing processes are applied for synthetic amorphous silica (SAS) production. These are the thermal process to obtain pyrogenic or fumed silica and the wet process to obtain hydrous silica, precipitated silica and silica gel (Younes et al. 2018b).

Owing to the hygroscopic structure of the amorphous silicon dioxide and its ability to absorb water; spray-dried materials, dry mixes or foods with high sugar content (juice powders, cocoa, coffee whiteners, etc.) prevent the negative effects of moisture (Villota et al. 1986). It is stated that silicon dioxide should be used in an amount not exceeding 2% by weight of host food powder.

It is also used as an adsorbent for substances such as dl-a-tocopheryl acetate and pantothenyl alcohol in various tableted foods. It must be used in the specified amounts to achieve the intended physical or technical effect. It can be also used as a stabilizer in the production of beer and is filtered from the beer before the last process step (Magnuson et al. 2013).

6.4 Stearates

Magnesium stearate E470 has a characteristic mild odor and feels oily when touched, it is a fine, light powder material that is practically white or close to white. It is insoluble in water and anhydrous ethanol. In addition to its use as an anticaking agent, magnesium stearate is also used as a lubricant and release agent, binder, thickener, emulsifier, and antifoaming agent. Magnesium stearate is used as an anticaking agent and also other functional purposes in food supplements (tablets, capsules, powders), herbs, spices, compressed and granulated mint sugar and confectionery, chewing gum, bakery products. It was determined that the maximum usage levels in these food categories ranged from 0.05 to 3% w/w (JECFA-CTA 2015). Magnesium stearate remains stable without decomposition in convenient storage conditions. It can absorb moisture during longer storage times (>12 months) (JECFA-CTA 2015).

Magnesium stearate is also used in cosmetics, pharmaceuticals, food, polymer, paper, rubber, paint industries. In these industries, it takes place as a gelling agent, stabilizer, lubricant, anti-adhesion, emulsifier, and plasticizer.

According to the study by Hobbs et al., Magnesium stearate did not show genotoxic effects. The study was concluded that the toxicity of magnesium stearate does not need to be assessed differently than other magnesium salts and confirms the acceptable daily intake (ADI) "not specified" of stearic and palmitic acids for magnesium salts (Hobbs et al. 2017).

Calcium stearate E470 is produced with the reaction of stearic acid and lime. It is a fine, white and silky touch powder. Calcium stearate remains stable under high-temperature conditions.

It is non-toxic and is used as a food additive thanks to its highly water-resistant and water-repellent properties. It acts also as a lubricant, stabilizer, and thickener in different industries. Calcium stearate is widely used as an anticaking agent and surface conditioner, especially in confectionery products (hard candies, tablet candies, etc.) (Lück et al. 2012).

Rebecca et al. published a study examining how calcium stearate and other anticaking agents affect the chemical and physical stability and moisture sorption of vitamin C. According to this study, anticaking agents increase the physical stability of the sodium ascorbate in powder form, while none of them have improved chemical stability of it (Lipasek et al. 2011).

6.4.1 Ferrocyanides

Sodium ferrocyanide E535, potassium ferrocyanide E536 and calcium ferrocyanide E538 were evaluated by JECFA and identified as food additives. The chemical formula of sodium, potassium, and calcium ferrocyanide are $Na_4[Fe(CN)_6]\cdot 10H_2O$, $K_4[Fe(CN)_6].3H_2O$ and $Ca_2[Fe(CN)_6].12H_2O$, respectively.

According to the specifications determined in the EU by Commission Regulation 231/2012 (EU) and JECFA (2006), purity of the content for food additive sodium, potassium, and calcium ferrocyanide should not be less than 99% by weight.

Hexacyanoferrate (II) anion $[Fe(CN)_6]_4$, which is commonly referred to as ferrocyanide, has a very stable structure due to the strong bonding between iron (+2 oxidation state) and each cyanide group. Free ferrocyanic acid; tetrahydrogen hexacyanoferrate ($H_4[Fe(CN)_6]$) is a strong tetrabasic acid when dissolved in water.

These three food additives, which are members of the ferrocyanides family, are completely synthetic. Sodium ferrocyanide (yellow prussiate of soda) is produced in an aqueous medium from crude sodium cyanide and iron sulfate. The sodium ferrocyanide decahydrate salt is recovered by crystallization. The potassium ferrocyanide is obtained by reacting sodium ferrocyanide with calcium hydroxide and potassium chloride. Calcium ferrocyanide is also obtained by reacting sodium ferrocyanide with calcium hydroxide (Younes et al. 2018a).

6.4.2 Talc E553b

According to FAO JECFA, Talc is in the form of hydrous magnesium silicate obtained from nature, containing various minerals such as alpha-quartz, calcite, chlorite, dolomite, kaolin, magnesite, and phlogopite. Talc, obtained from deposits with asbestos content, is not suitable for use in food-grade because of the carcinogenic effect of asbestos. It is also described as a fine, odorless, white or greyish white crystalline powder, adheres easily to the skin, and is free from grittiness (JECFA 2019).

6.5 Phosphates

Tricalcium phosphate E341 (iii) is a calcium salt of phosphoric acid. It is commercially produced from phosphoric acid, which is obtained from a phosphate mine. The acceptable daily intake was determined as 70 mg/kg body weight. Phosphoric acid and phosphates do not have any dietary restrictions and they are normally suitable for consumption by groups, vegans and vegetarians (FDA 2019).

Trimagnesium phosphate E340 (iii) is described as a white, odorless crystalline powder. The chemical formula is $Mg_3(PO_4)_2$ and it also has various hydrates form (JECFA 2019).

6.6 Novel Anticaking Agents

In recent years, efforts have been made to develop innovative food additives that solve many problems at the same time. The health status of food consumers (food allergies etc.), dietary preferences (vegan nutrition, etc.) and religious restrictions also affect food additives preferences. For example, many companies prefer non-animal sources for the production of stearate products.

Silica aerogel products can be given as another example to novel anticaking agents. Silica aerogels have superior features and it provides many advantages for use as an anticaking agent in food powders. High surface area, low bulk density, very fine particle size are among its prominent features. Moisture adsorption capacity is really high compared to traditional anticaking agents which are the same weight. Thus, even small amounts of silica aerogel can be used in food powders effectively (Dorcheh and Abbasi 2008; Yücel et al. 2016; Temel et al. 2017).

Another approach is to offer alternative products instead of classical food additives, the use of which is controversial in terms of safety. Geertman (2005) evaluated the use of new generation anticaking agents, metal-organic complexes instead of classical anticaking agents (ferrocyanide), which are frequently used to regulate the flow of salts. Metal ions prevent oxidation by forming an oxide layer on the surface of the salt (Geertman 2005). Bode et al. examined the anticaking effect of Iron (III) meso-tartrate, which is also a metal-organic complex, on sodium chloride salt and explained its anticaking activity (Jiang et al. 2016).

6.7 Dietary Exposure to Anticaking Agents

The food additives status list provides brief indications about the usage limitations for each additive. Special regulations prepared for each item should be examined to get detailed information about the limitations of use (FDA 2019).

There are strict regulations and limits on the use of preventive substances in food powders. Most importantly, the preventive substances should be inert, safe in the defined quantities and be specified as "Generally Recognized as Safe" (GRAS). Besides, it is legally permitted to use 2% or less of host powder. For this reason, it should be effective even at low concentrations. Various experiments are carried out on the host powder to find the appropriate concentration and the most effective working anticaking agent (Aguilera 2005; Lipasek et al. 2011)

It is very important in terms of safety to determine the appropriate dose in foods. Acceptable Daily Intake (ADI) is defined as milligrams per kilogram of body weight per day (mg/kg bw/day) and is widely used by national/international regulatory and advisory committees working worldwide. To determine an ADI, biological and toxicological studies (cytotoxicity, genotoxicity, carcinogenicity, etc.) are evaluated. These studies determine the dose levels of the additive that can cause adverse effects on *in vivo* test animals (usually rats or mice) health and *in vitro* cell lines. Another significant term for food additives, maximum usage level is expressed as the highest concentration of an additive that is determined to be functionally effective in a food or food category and is considered safe and is commonly referred to as mg additive/kg food (CODEX 2019). ADI values and maximum usage levels of commonly used anti-caking agents are shown in the Table 6.1.

6.8 Regulations

Standardization of food production safety and safe food consumption is an important issue all over the world. The two main regulatory foundations of food additives that are recognized worldwide are the European Food Safety Authority (EFSA) and the United States Food and Drug Administration (FDA). Rules, definitions, technical information, usage amounts, labeling and procedures related to food additives are determined by prepared regulations. In order to re-evaluate the safety of food additives, EFSA makes open calls in which scientific opinions are provided at certain times. For example, 115 scientific opinions published by EFSA for the re-evaluation of safety for 204 of 316 food additives approved before January 20 2009. For the remaining 112 food additives, calls were made to be re-evaluated by EFSA before 31 December 2020. United Nations Food and Agriculture Organization (FAO). JECFA performs risk assessments on food additives and serves as an independent scientific committee to WHO and FAO organizations, member countries and these organizations.

Table 6.1 Anticaking agents as food additives with usage and their respective ADI quantities (mg/kg bw) and maximum usage levels (mg/kg)

E number	Name	ADI	Maximum usage level	Food group	References
E551	Silicon dioxide	Not specified	2000–30,000 mg/kg	22 food categories	Younes et al. (2018a)
E552	Calcium silicate	Not specified	5400–30,000 mg/kg	13 food categories	Younes et al. (2018c)
E553a	Magnesium silicate	Not specified	5400–30,000 mg/kg	13 food categories	Younes et al. (2018c)
E553b	Talc	Not specified	5400–30,000 mg/kg	13 food categories	Younes et al. (2018c)
E170	Calcium carbonate	Not specified	6000 mg/kg	Baking powder	Opinion (2011)
E470	Magnesium stearate	Not specified	20,000 mg/kg	Chewing gums	JECFA-CTA (2015)

Food additives were collected under 26 functional classes according to Regulation (EC) No 1333/2008. Class names and the international numbering system are very important and necessary arrangement for food additives. The International Numbering System (INS) for Food Additives is a European-based labeling system for food additives that aims to provide a brief description of materials that can have a long real name. INS numbers consist of three or four digits. In the European Union (EU), approved food additives are written with the prefix E (E represents Europe). Countries outside Europe can use the number system without an E suffix. With this label, all food additives have been determined whether approved or not for use in foods (Carocho et al. 2014; CAC 2019).

A food additive can be used to provide solutions to different problems within the food, and it is the manufacturer's responsibility to declare the most important functional class in the list of ingredients. For example, calcium carbonate can act as a surface colorant, stabilizer or anticaking agent in foods and is therefore appropriately marked as "anticaking agent INS 170" or "surface colorant 170" in the ingredients list (CAC 2019).

References

Aguilera, J. M. (2005). *Food powders: Physical properties, processing, and functionality.* New York: Kluwer Academic. https://doi.org/10.1007/0-387-27613-0.

Aguilera, J. M., Valle, J. M., & Ka, M. (1995). Caking phenomena in amorphous food powders. *Trends in Food Science and Technology, 6*, 149.

Branen, A. L., Davidson, P. M., & Salminen, S. (2002). *Food additives* (2nd ed.). Boca Raton: CRC Press. Revised and expanded.

CAC (2019). *Class names and the international numbering system for food additives.* Codex Alimentarius. http://www.fao.org/fao-who-codexalimentarius/sh-proxy/en/?lnk=1&url=https

%253A%252F%252Fworkspace.fao.org%252Fsites%252Fcodex%252FStandards%252FCXG%2B36-1989%252FCXG_036e.pdf.

Carocho, M., Barreiro, M. F., Morales, P., & Ferreira, I. C. F. R. (2014). Adding molecules to food, pros and cons: A review on synthetic and natural food additives. *Comprehensive Reviews in Food Science and Food Safety, 13*(4), 377–399. https://doi.org/10.1111/1541-4337.12065.

CODEX (2019). *General standard for food additives*. http://www.fao.org/gsfaonline/docs/CXS_192e.pdf. Access date: 15.02.2019.

Chang, L. S., Karim, R., Abdulkarim, S. M., Yusof, Y. A., & Ghazali, H. M. (2018). Storage stability, color kinetics and morphology of spray-dried soursop (Annona muricata L.) powder: Effect of anticaking agents. *International Journal of Food Properties, 21*(1), 1937–1954. https://doi.org/10.1080/10942912.2018.1510836.

Dorcheh, A. S., & Abbasi, M. H. (2008). Silica aerogel; synthesis, properties and characterization. *Journal of Materials Processing Technology, 199*(1–3), 10–26. https://doi.org/10.1016/j.jmatprotec.2007.10.060.

EFSA. (2012). Scientific opinion on the safety and efficacy of bentonite as a technological feed additive for all species. *EFSA Journal, 10*(7), 2787. https://doi.org/10.2903/j.efsa.2012.2787.

Emerton, V., Choi, E., House, T. G., Park, S., & Road, M. (2008). *Essential guide to food additives*. Cambridge: Royal Society of Chemistry.

Ermiş, E., Güneş, R., & Zent, İ. (2018). Bazı Model Toz Gıdaların Akışkanlığına ve Sıkıştırılabilirliğine Partikül Boyutunun Etkisinin PFT Toz Akışı Test Cihazı Kullanılarak Belirlenmesi. *Türk Tarım – Gıda Bilim ve Teknoloji Dergisi, 6*(1), 55–60.

FDA (2019). Food additives permitted for direct addition to food for human consumption. Subpart E--Anticaking Agents. CFR - Code of Federal Regulations, Sec. 172.430 Iron ammonium citrate. Revised as of April 1, 2019.

Ganesan, V., Rosentrater, K. A., & Muthukumarappan, K. (2008). Flowability and handling characteristics of bulk solids and powders – A review with implications for DDGS. *Biosystems Engineering, 101*(4), 425–435. https://doi.org/10.1016/j.biosystemseng.2008.09.008.

Geertman, R. M. (2005). How to make salt rust or: New anticaking agents for salt. VDI-Gesellschaft Verfahrenstechnik und Chemieingenieurwesen. In: Industrial crystallization, 2, 557–562. ISBN: 3180919019

Hansen, L. D., Hoffmann, F., & Strathdee, G. (1998). Effects of anticaking agents on the thermodynamics and kinetics of water sorption by potash fertilizers. *Powder Technology, 98*(1), 79–82. https://doi.org/10.1016/s0032-5910(98)00037-0.

Hobbs, C. A., Saigo, K., Koyanagi, M., & Hayashi, S.-M. (2017). Magnesium stearate, a widely-used food additive, exhibits a lack of in vitro and in vivo genotoxic potential. *Toxicology Reports, 4*, 554–559. Retrieved from https://pubmed.ncbi.nlm.nih.gov/29090120.

Hollenbach, A. M., Peleg, M., & Rufner, R. (1982). Effect of four anticaking agents on the bulk characteristics of ground sugar. *Journal of Food Science, 47*(2), 538–544. https://doi.org/10.1111/j.1365-2621.1982.tb10119.x.

Irani, R. R., Callis, C. F., & Liu, T. (1959). Flow conditioning anticaking agents. *Industrial and Engineering Chemistry, 51*(10), 1285–1288. https://doi.org/10.1021/ie50598a035.

Jiang, S., Meijer, J. A. M., Van Enckevort, J. P., & Vlieg, E. (2016). Structure and activity of the anticaking agent iron (III) meso-tartrate. *Dalton Transactions, 45*, 6650.

JECFA-CTA. (2015). Chemical anf Technical Assessment (CTA). http://www.fao.org/food/food-safety-quality/scientific-advice/jecfa/technical-assessments/en/. Access date: 09.01.2019.

JECFA. (2019). Evaluations of the Joint FAO/WHO Expert Committee on Food Additives. https://apps.who.int/food-additives-contaminants-jecfa-database/search.aspx. Access date: 30.03.2019.

Juliano, P., & Barbosa-Canovas, G. V. (2010). Food powders flowability characterization: Theory, methods, and applications. *Annual Review of Food Science and Technology, 1*, 211–239.

Lipasek, R. A., Taylor, L. S., & Mauer, L. J. (2011). Effects of anticaking agents and relative humidity on the physical and chemical stability of powdered vitamin C. *Journal of Food Science, 76*(7), C1062–C1074. https://doi.org/10.1111/j.1750-3841.2011.02333.x.

Lück, E., Aktiengesellschaft, H., & Republic, F. (2012). Foods, 3. Food additives. In *Ullmann's encyclopedia of industrial chemistry* (pp. 671–691). Weinheim: Wiley-VCH.

Martinez, J. A. R., & Rocafull, M. (2015). *Anti-caking compositions for fertilizers*. U.S. Patent No. US8932490B2.

Magnuson, B., Munro, I., Abbot, P., Baldwin, N., Lopez-Garcia, R., Ly, K., & Socolovsky, S. (2013). Review of the regulation and safety assessment of food substances in various countries and jurisdictions. *Food additives & contaminants: Part A, 30*(7), 1147–1220.

McNutt, K., & Sentko, A. (2003). Isomalt. In *Encyclopedia of food sciences and nutrition* (pp. 3401–3408). New York: Academic Press. https://doi.org/10.1016/b0-12-227055-x/00658-1.

Meiron, O. E., Bar-David, E., Aflalo, E. D., Shechter, A., Stepensky, D., Berman, A., & Sagi, A. (2010). Solubility and bioavailability of stabilized amorphous calcium carbonate. *Journal of Bone and Mineral Research, 26*(2), 364–372. https://doi.org/10.1002/jbmr.196.

NCBI (2020a). PubChem Compound Summary for CID 10176262, Calcium bicarbonate. National Center for Biotechnology Information. https://pubchem.ncbi.nlm.nih.gov/compound/Calcium-bicarbonate.

NCBI (2020b). PubChem Compound Summary for CID 129627764, Calcium hydrogen carbonate. National Center for Biotechnology Information. https://pubchem.ncbi.nlm.nih.gov/compound/129627764.

NCBI (2020c). PubChem Compound Summary for CID 24456, Calcium phosphate. National Center for Biotechnology Information. https://pubchem.ncbi.nlm.nih.gov/compound/Calciumphosphate.

Opinion, S. (2011). Scientific opinion on re-evaluation of calcium carbonate (E 170) as a food. *EFSA Journal, 9*(7), 1–73.

Pandey, R. M., & Upadhyay, S. K. (2012). *Food additive*. London: IntechOpen. https://doi.org/10.5772/34455.

Rashid, I., Daraghmeh, N. H., Al Omari, M. M., Chowdhry, B. Z., Leharne, S. A., Hodali, H. A., & Badwan, A. A. (2011). Magnesium silicate. In *Profiles of drug substances, excipients and related methodology* (pp. 241–285). San Diego: Elsevier. https://doi.org/10.1016/b978-0-12-387667-6.00007-5.

Rychen, G., Aquilina, G., Azimonti, G., Bampidis, V., Bastos, M. D. L., Bories, G., et al. (2017). Safety and efficacy of bentonite as a feed additive for all animal species. *EFSA Journal, 14*, 15.

Taylor, P., Slade, L., Levine, H., Reid, D. S., Slade, L., & Levine, H. (1991). Beyond water activity: Recent advances based on an alternative approach to the assessment of food quality and safety. *Critical Reviews in Food Science and Nutrition, 30*(2–3), 115–360.

Temel, T. M., İkizler, B. K., Terzioğlu, P., Yücel, S., & Elalmış, Y. B. (2017). The effect of process variables on the properties of nanoporous silica aerogels: An approach to prepare silica aerogels from biosilica. *Journal of Sol-Gel Science and Technology, 84*(1), 51–59. https://doi.org/10.1007/s10971-017-4469-x.

Villota, R., Hawkes, J. G., & Cochrane, H. (1986). Food applications and the toxicological and nutritional implications of amorphous silicon dioxide. *CRC Critical Reviews in Food Science and Nutrition, 23*(4), 289–321. https://doi.org/10.1080/10408398609527428.

WHO (2006). Enhancing Developing Country Participation in FAO/WHO Scientific Advice Activities: Report of a Joint FAO/WHO Meeting, Belgrade, Serbia and Montenegro, 12–15 December 2005 (Vol. 88)

Younes, M., Aggett, P., Aguilar, F., Crebelli, R., Dusemund, B., Filipič, M., Frutos, M. J., Galtier, P., Gott, D., Gundert-Remy, U., Kuhnle, G. G., Lambré, C., et al. (2018a). Re-evaluation of sodium ferrocyanide (E 535), potassium ferrocyanide (E 536) and calcium ferrocyanide (E 538) as food additives. *EFSA Journal, 16*(7), e05374. https://doi.org/10.2903/j.efsa.2018.5374.

Younes, M., Aggett, P., Aguilar, F., Crebelli, R., Dusemund, B., Filipič, M., Frutos, M. J., Galtier, P., Gott, D., Gundert-Remy, U., Kuhnle, G. G., Leblanc, J., et al. (2018b). Re-evaluation of silicon dioxide (E 551) as a food additive. *EFSA Journal, 16*(1), e05088. https://doi.org/10.2903/j.efsa.2018.5088.

Younes, M., Aggett, P., Aguilar, F., Crebelli, R., Dusemund, B., Frutos, M. J., et al. (2018c). Re-evaluation of calcium silicate (E 552), magnesium silicate (E 553a (i)), magnesium trisilicate (E 553a (ii)) and talc (E 553b) as food additives. *EFSA Journal, 16*, e05375.

Yücel, S., Karakuzu İkizler, B., & Temel, T. M. (2016). Aerojel: Üstün Özellikleri, Çeşitleri ve Gelişen Uygulama Alanları. *Turkchem*, pp. 52–64.

Zafar, U., Vivacqua, V., Calvert, G., Ghadiri, M., & Cleaver, J. A. S. (2017). A review of bulk powder caking. *Powder Technology, 313*, 389.

Chapter 7
Modification of Food Powders

Nasim Kian-Pour, Duygu Ozmen, and Omer Said Toker

7.1 Introduction

Food powders constitute an important part of the food industry and they are used as both raw materials/ingredients (hydrocolloids, flour, starch etc.) and processed products (instant coffee, powdered fruits, honey powders etc.). They are used in the food products for many purposes which are determined considering their composition, microstructure, chemical and physical properties. Food powders are modified to improve physicochemical characteristics. Decreasing the particle size of the sugar particles can provide an opportunity to decrease the sugar content of the products, which is important for the production of low-calorie products. In this way, the surface area of the sugar is increased, which results in increasing contact points between sugar particles and receptors. Therefore, a similar sugar taste can be perceived by a lower amount of sugar with a reduced particle size. Nestle announced a patent which is related to the improvement of solubility of sugar and in this way they decreased the sugar content of the chocolate up to 40%. In addition, density, compressibility, fluidity, solubility, hydration and surface properties of the food powders play a crucial role in obtaining the products with desired characteristics. These properties are also improved by modification of the powders. Agglomeration is a widely used process for modification of the powders.

Size enlargement is a term that includes a variety of processes incorporating small particles into larger stable masses by means of various techniques, in which the original particles are still detectable. Size enlargement operations are used by a wide variety of industries to obtain a broad range of benefits providing results such

N. Kian-Pour (✉)
Department of Food Technology, School of Applied Science, Istanbul Aydin University, Istanbul, Turkey

D. Ozmen · O. S. Toker
Department of Food Engineering, Faculty of Chemical and Metallurgical Engineering, Yildiz Technical University, Istanbul, Turkey

E. Ermiş (ed.), *Food Powders Properties and Characterization*, Food Engineering Series, https://doi.org/10.1007/978-3-030-48908-3_7

as lowering dusting or product losses, reducing powder dispersion in the environment and lowering inhalation, reducing hazardous of handling harmful and toxic chemicals and wastes, giving powders free-flowing, densifying products for better transport and storage, decreasing caking and lump formation, producing structural forms, enhancing the appearance of powders, controlling of powder properties (porosity, heat transfer rates, solubility), obtaining uniform mixtures of solids which do not separate, and more homogeny distributing the active molecules. Therefore, size enlargement processes received a great interest in the food, pharmaceuticals, detergents, agricultural, nutraceuticals, cosmetics, and mineral processing (Reid 1974).

Size enlargement processes include briquetting, pelletizing, tableting, and agglomeration. Briquetting is the densification or compaction of residue and biomass waste which produce briquetted products with a higher density than the raw materials. Briquetted products can be used as a source of energy in households or large industries. The coffee husks, rice bran, maize cobs, rice husks, gum Arabica, and tree leaves are the raw materials in producing the fuel briquettes (Kundu et al. 2017). Pelletizing operations press moist single ingredients or mixture through die openings and cutting off the rod like products into extruded pellets forms. In principle, heat, moisture and pressure applied by extruders to fine and difficult–to–handle materials to agglomerate them into the larger particles with better handling properties such as pharmaceuticals, foodstuffs (snack foods), fertilizer products and animal feeds (Aguilar-Palazuelos et al. 2012). Tableting is one of the size enlargement processes which compressed a powder or granule mixture to the compressed agglomerate product which used in the pharmaceutical, cosmetic, dietary supplement, catalysts, fertilizers, pesticides, cleaning agents, ceramics, candies, sweeteners, stock cubes, salt tablets and sugar industry.

Agglomeration is known as an important size enlargement operation during which small particles are joined together to construct bigger particles. The main agglomeration techniques are pressure agglomeration, growth agglomeration (tumble/agitated methods), agglomeration by spray methods. Non–agglomeration processes which improve the properties of food powder are freeze drying, thermal treatment of amorphous foods, osmotic and drum drying, separate fat from products, adding the porogens or templating agents and removing them by different techniques to produce porous particles with low density and open or closed pore structures (Saravacos and Kostaropoulos 2002). Agglomeration is used by various industries to improve the characteristics of the agglomerated products compared to the traditional non-agglomerated products such as flow properties, safer and cheaper transport and storage, and easy to use by the consumers. It can produce a non-dusting powder like pesticides and detergents and reduce the hazards of spreading them in the environment during transport. Besides, the problems related to the low bulk densities and difficult flow characteristics of powders during handling and storage can be solved by agglomeration by increasing bulk densities of products and producing a more free flowing nature to decrease segregation of individual particles from each other and held them together in the formed agglomerate (Ennis 1996). Agglomeration is used in many food and non-food industry such as detergents, fertilizers, mineral and clay products, mammal feeds, ceramics, enzyme and yeast, pharmaceutical products, baking powder, ready-to-cook mixtures, beverage powder, flow table salt, pudding powder, spices, ready-to-eat soups, compacted cubes of soups, chocolate and dispersible milk powder industry. Agglomeration improves bulk density, followability, and control the porosity of products (Dhanalakshmi et al. 2011). Also, it can create products with rapidly dispersion properties in a liquid such as instant milk, chocolate, coffee, cocoa, soft drinks, sugar mixture, soups, flours, starches, vitamins, dextrin, and drug powders. The instantaneous properties of agglomerates can be measured by wettability, sinkability, dispersibility and solubility of the products which are wetted homogeneity in hot or cold water. The size of agglomerates is varied from 0.1 to 3 mm (Barbosa-Canovas et al. 2005). In this chapter modification methods of food powders are summarized and their effects on the quality characteristics of the food powders will be mentioned.

7.2 Agglomeration

Agglomeration is the process in which solid particles sticking together in a random way by different physical or chemical forces and create a larger aggregate of a porous and extended structure while their single shapes remain unchanged. The agglomerated products need to be strong enough to withstand during handling and storage and need to disperse easily in liquids (Saravacos and Kostaropoulos 2002). Agglomerated food products can use directly as the final products by consumer (e.g., milk powder, baby foods, coffee, beverage powder, vitamins and minerals, sweeteners, salt, sugar, onion and garlic powder), or indirectly as the products which are used by further food processing (e.g., starch, flour, egg powder, gums, yeast, enzymes). Furthermore, agglomerated food powders can be used as a coating material (maltodextrin), flavoring (cheese powder, spices), or as a material to aid for drying (starches/flour) (Dhanalakshmi et al. 2011).

The agglomeration used different unit operations such as spraying, heating, drying, steaming, pressurizing, agitating, extruding, etc., aimed at agglomerating particles. The technologies applied varied depends on different factors such as particle size, thermal sensibility, process conditions, physical and chemical properties of products, and adhesion principles Dhanalakshmi et al. 2011). Processes used for agglomeration can be divided mainly into (1) pressure methods (i.e., extrusion), (2) tumble/growth and agitated methods (i.e., inclined rotating drum), (3) thermal processes (i.e., steam jet and drum dryer), (4) spray techniques (i.e., spray dryer). Also, agglomeration can be divided into "wet" and "dry" according to usage of a binder liquid in the process which generally wet methods are referred to as granulation (Ennis 1996; Green 2007).

7.2.1 Bonding and Adhesion Forces Between Particles

The strength of an agglomerate mainly depends on adhesion forces between particles. For successful agglomeration processes, the strength of the bonds between particles must be higher than disruptive forces to prevent the breakdown of the agglomerated products during handling operations. However, the magnitude of these forces depends on the size, structure and moisture content of particles (Green 2007). Four major bonding mechanisms involved in agglomeration processes but more than one may apply during the process of agglomeration (Fig. 7.1). They are liquid bridges, solid bridges, intermolecular and intramolecular forces, and mechanical interlocking (Barbosa-Canovas et al. 2005).

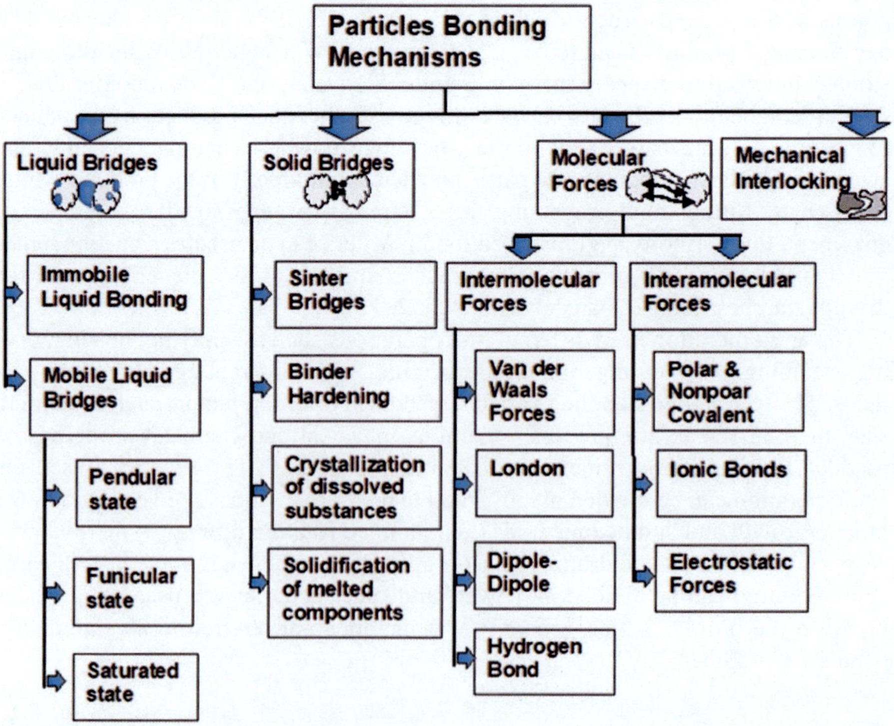

Fig. 7.1 Bonding mechanisms between powder particles [adapted from Barbosa-Canovas et al. 2005]

7.2.1.1 Liquid Bridges

In the immobile liquid bridges, the adhesion forces (between a particle and surface of a material) and/or cohesion (between particles of the same material) forces by adsorption layers and highly viscous component contribute in forming bonds between particles by the following mechanisms (Buffo et al. 2002):

- When sufficient water is available, a thin and immobile adsorption layer appears on the surface of fine particles. This film decreases the distance and increases the connection zones between particles and causes to the formation of liquid bridges.
- In high viscous materials, the thin immobile film generates strong bonds that are more powerful than bonds formed by mobile liquid layers.

The mobile liquid bridges due to the interfacial forces (surface tension) and capillary power (suction and pressure) have an important role to hold the particles together during the agglomeration process which can be obtained at three states (Green 2007; Simons 2007):

- The pendular state in which capillary forces attract all free moisture to the junctions between particles and surface tension pulls the particle together which forms a lens–shaped arc at the contact points of particles.
- With an increase in the liquid content, all internal particle surfaces become restricted by liquids, and a continuous network of liquid contact with air and the lens–shape rings coalesce gives rise to the funicular state.
- At the fully saturated state, all pores are completely filled with liquids and the agglomerated products reach their capillary state.

7.2.1.2 Solid Bridges

The deposited materials between the powders can form solid bridges. They might have developed by different mechanisms such as sinter bridges, hardening of binders, crystallization of dissolved substances, chemical bonds, and solidification of melted components (Buffo et al. 2002).

- When high pressures and temperatures are used, at the contact points between the particles, partial melting is taking place, which causes the molecules to diffuse from one particle to the adjacent particle and create solid bridges.
- The amorphous foods at elevated humidity or temperature (below the melting temperature of the powder), form solid bridges between particles which are related to growing sinter bridges.
- The drying operation in the agglomeration process can increase the solid content inside the liquid bridges between crystalline particles, therefore viscous forces are increased and the viscoelastic bridge is obtained which solidifies viscous bridges by drying constitute the solid bridges.
- Capillary condensation is another way for the formation of solid bridges. During compression of powders, capillary condensation may take place which releases

moisture at the contact points of particles and produce a liquid bridge between them. This water can dissolve crystalline substance and if this water evaporates during storage or drying, dissolve material recrystallizes and create solid bridges between neighboring particles.

- The solid bridges can be formed by surface crystallization of dissolved material during dehydration.
- In the fatty food powders solid bridges can form by melting and recrystallization of fat by heating the powder near to melting temperature following the cooling which is named melt agglomeration.
- The binder materials used in the agglomeration process (such as starch solution) can act as a glue to stick particles together which during drying operation create solid bridges.

7.2.1.3 Intermolecular and Intramolecular Forces

Very fine particles can join to each other without the presence of liquid or solid bridge via intermolecular and intramolecular (electrostatic) forces. These short–range forces can be effective for the particles of less than 1 micron diameter (Feng and Hays 2003; Green 2007).

- Intermolecular forces represent the attraction and repulsion forces that exist between molecules as a group known as van der Waals forces which are related to the different polarization mechanisms and consist of dispersion (London), dipole–dipole, and hydrogen bonds. The van der Waals forces can be effective when particles are very close to each other. This limitation causes the overall power of van der Waals forces on powder particles to display high sensitivity to the surface structures at the microscopic level.
- Intramolecular forces exist between the two oppositely charged ions and occur within molecules/substances (polar and nonpolar covalent bonds, ionic bonds or electrostatic forces). The interchange of electrons and ions from the surface of one particle to another particle is the basic of electrostatic forces between solid particles. The powder particles with a significant amount of net electric charge can bond to each other by electrostatic forces. However, in the lack of significant charge on powder particles, van der Waals forces are dominant to electrostatic forces. It is possible to give an electric charge to powder by contacting with other substances or by use of an externally applied electric field.

7.2.1.4 Mechanical Interlocking

Form-related bonds or the mechanical interlocking of fibrous particles or plaited materials during the mixing or compression operations can create "form–closed" bonds. Although it is generally has a little contribution in the agglomeration strength in comparison with the other binding forces.

7.2.2 Agglomeration Strength

Mechanical strength of agglomerated powder is an important property in the possibilities of subsequent processing and depends on all forces and bonds which hold particle agglomerate together. However, because of its complexity, theoretical models are developed to determine the strength of agglomerated particles. The tensile strength for regular packs of monosized spherical particles can be described by general Eq. 7.1 (Green 2007):

$$ts = \left(\frac{9}{8}\right)\left[\frac{1-\varepsilon}{\pi d^2}\right]kF \tag{7.1}$$

where *ts, d, ε, k* and *F* are tensile strength (kg/cm^2), the particle diameter (cm), the void or porosity volume fraction, the coordination number (average number of contact points of one particle with adjacent particles), and the bonding force at the contact point of particles caused by binding mechanisms (kg/kg), respectively. The coordination number can be calculated from Eq. 7.2 (Tsubaki and Jimbo 1984).

$$k = \frac{\pi}{\varepsilon} \tag{7.2}$$

Generally, increase in the particle size and decrease in the interparticle distance causes to increase in all binding forces. When the distance between particles reaches to 1 μm or higher, van der Waals forces are practically zero while the distance between particles is not so critical for the strength of liquid bridges. And finally, with an increase of interparticle distance more than 1 μm, electrostatic forces are the main force to bond particles together (Barbosa-Canovas et al. 2005). The agglomeration strength after the creation of agglomerated material, in the availability of moisture, depends on liquid bridges; otherwise, it depends on van der Waals forces. Besides, in wet agglomeration viscous forces are responsible for the agglomerate strength (Knight 2001). Furthermore, the strength of agglomerate structures has an inverse relation with porosity. However, porosity can be minimized by correct particle size distribution. In liquid bridges:

- For pendular state tensile strength can be calculated from Eq. 7.3 (Green 2007):

$$ts = 2.8\left(\frac{1-\varepsilon}{\varepsilon}\right)\frac{\sigma}{d\,f(\delta)} \tag{7.3}$$

where σ represents the surface tension of binder material (N/cm) and δ is the contact angle (rad). In this situation, the tensile strength of agglomerated material is about $\frac{1}{3}$ of tensile strength at capillary state while tensile strength funicular condition is between pendular and capillary state.

- Tensile strength at capillary state can be determined by Eq. 7.4:

$$ts = 8.0\left(\frac{1-\varepsilon}{\varepsilon}\right)\frac{\sigma}{d\,f(\delta)} \tag{7.4}$$

- In the agglomerate structure when wetting is complete and the solid is completely filled by liquid, the value of $f(\delta) = 1$ and tensile strengh are expressed as Eq. 7.5.

$$ts = c\left(\frac{1-\varepsilon}{\varepsilon}\right)\frac{\sigma}{x_{sv}} \tag{7.5}$$

where x_{sv} is the surface equivalent diameter (surface volume diameter) of the particle (Barbosa-Canovas et al. 2005).

- For agglomeration of non–metallic powder by compact pressure methods, Eq. 7.6 can be used:

$$\log p = m\frac{V}{V_s} + b \tag{7.6}$$

where, p, is pressure applied to compact, m and b are constants, V is the volume of the compact at the applied pressure and V_s represents the volume of solid powder (void-free) (Green 2007).

7.2.3 Binders

In most food agglomerates, binders are necessary to improve the agglomerates strength. Binders are liquid or dry materials and their adhesive properties provide the necessary cohesiveness and capillary and viscous forces to bond solid particle together. At the agitated agglomeration process, the rate of size enlargement and the size distribution are mainly affected by binder viscosity (Knight 2001). The binder can be mixed at the dry state with powder and agglomerating solvent (generally water) or firstly dissolve in the solvent then added to the powder. Commonly, a liquid binder is spread on powder by pump or atomized during mixing operation and agglomerates growth due to different mechanisms such as nucleation, coalescence and layering. After evaporation of the solvent, the particles stick together and form a large agglomerate. The mixer provides shearing forces that consolidate the powder and with the final solidification and drying operations, strong agglomerates are forms (Tardos et al. 1997). However, in several cases, a combination of binders and other materials in the formulation of agglomerates are used such as flow aids, flavor and taste modifiers, wetting agents, emulsifiers, antioxidants, edible colors, surfactants and materials which produce CO_2 (beverages). In most cases, water is used as a binder in a large extent in the food industry. The moisture content of agglomerated powders has a very important impact on agglomerate's quality. Agglomeration processes at high moisture content and longtime produce porous granules while low moisture content and longtime process, lead to the formation of high-density granules. High water content has a positive impact on the increase in the size and porosity of agglomerated materials (Saravacos and Kostaropoulos 2002).

In agglomeration by very low viscosity binders, the surface tension forces are dominant when comparing with viscous forces and every agglomerate has its specific optimal moisture content. The binder wetting properties are correlated with surface tension and contact angle. When contact angle is close to a critical angle 90°, the wetting characteristics become critical and for contact angels over the 90°, agglomerates show the poor characteristics such as wide range of size distribution and very low strength. Although, to modify the wetting ability incorporation of a surface active compound is recommended (Knight 2001). Besides water, various binders can be used in the food industry which is offered in Table 7.1 (Barbosa-Canovas et al. 2005).

The physical and mechanical properties of a binder materials such as concentration, viscosity, cohesion, adhesion, wettability, binder–particle interactions, film–forming properties and the distribution of the binder in the agglomerates, have an important effect on the binder efficiency. For example, a binder with high viscosity (starch paste), can produce more brittle agglomerate, gelatin or acacia gum can form agglomerates with high hardness, sucrose create hard and brittle bridges (Barbosa-Canovas et al. 2005).

Table 7.1 Binders used in agglomeration [Adapted from Augsburger and Hoag 2008]

Binders	Agglomeration technique	Formulation percentage (%)
Starch	Wet mixing	2–5
Pre-gelatinized starch	Wet mixing	2–5
Pre-gelatinized starch	Dry mixing	5–10
Sodium alginate	Wet mixing	1–3
Gelatin	Wet mixing	1–3
Alginic acid	Dry mixing	3–5
Methylcellulose	Wet mixing	1–5
Methylcellulose	Dry mixing	5–10
Na-Carboxymethylcellulose	Wet mixing	1–5
Na-Carboxymethylcellulose	Dry mixing	5–10
Sucrose	Wet mixing	2–25
Glucose	Wet mixing	2–25
Sorbitol	Wet mixing	2–10

7.3 Agglomeration Techniques

The three basic methods of agglomeration are (1) pressure agglomeration (i.e., extrusion), (2) growth agglomeration (tumble/agitated methods), (3) spray techniques (i.e., spray dryer). Also, there are alternative agglomeration processes such as steam granulation, thermal adhesion agglomeration, and freeze agglomeration which was shown in Fig. 7.2 (Green 2007; Shanmugam 2015).

7.3.1 Pressure Agglomeration

In compaction or pressure, agglomeration pressure forces are applied to the small particles system in a limited space which is then shaped and densified therefore creates larger cohesive agglomerates (Barbosa-Canovas et al. 2005). Examples of dry agglomeration techniques are roll compaction and uniaxial die compaction. As pressure dry agglomeration carry outs in dry condition without a need for liquid binders and drying processes, it is a cost effective method and also it is a suitable method for agglomeration of moist sensitive materials (Dhanalakshmi et al. 2011). It has a wide application in chocolate, sugar, candies, flakes, pellets and pasta pro-

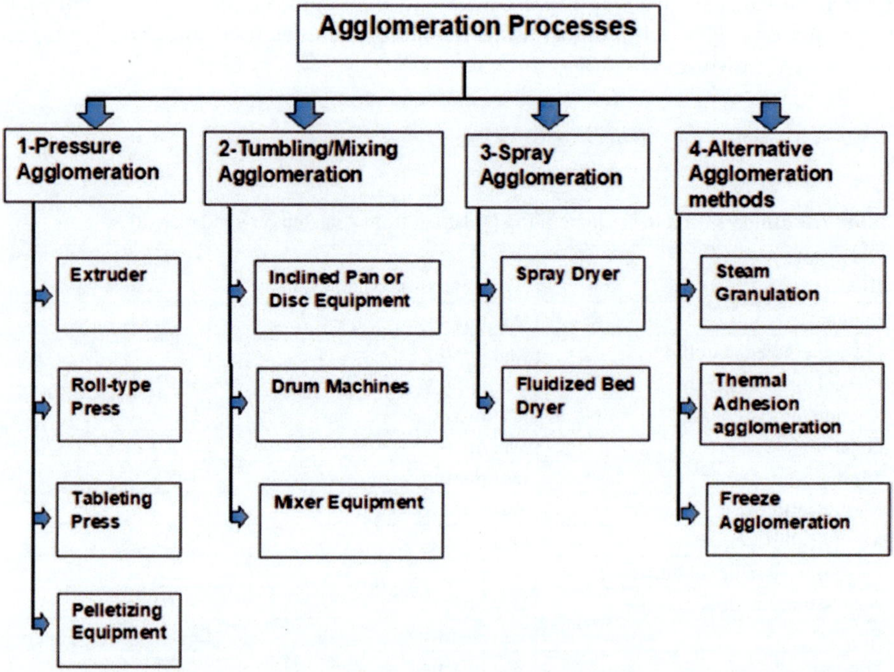

Fig. 7.2 Agglomeration Process and equipment [adapted from Green 2007]

cessing. Pressure methods generally carried out in two steps; in the first step applied pressure creates a forced rearrangement of particles and particles fill the large holes, while in the second step pressure rises sharply and causes the fragmentation of brittle particles and plastic deformation of soft particles which fill the smaller holes (Popescu and Vidu 2018).

Factors influencing compression or compaction agglomeration process are:

1. Raw material properties (form, size, structure, stickiness, ability to form inter particle bonds during compression or consolidation, moisture content).
2. Effective utilization and transfer of the applied pressure power.
3. Duration of compaction or compression.
4. The temperature of powder during operation.

The different advantages and disadvantages of pressure agglomeration are illustrated in Table 7.2.

However, elastic springback and compressed air in the pores are two main reasons to limit the speed of compaction and processing capacity. These two phenomena can create a crack in the products and decrease the strength of agglomerated products. One way to reduce these effects is by maintaining the maximum pressure for some times (dwell time), prior to release pressure. Compaction agglomeration is carried out in various equipment like as piston or molding presses, tableting presses, roll presses, extruders, and pellet mills (Green 2007; Barbosa-Canovas et al. 2005).

Pressure agglomeration can be done at different pressure levels:

- Low and medium pressure mode is characterized by uniformity of particle size, with generally elongated spaghetti–like or cylindrical shape products. Generally, sticky mixtures of fine particles and binders forced through holes, perforated dies and differently shaped screen therefore pressure and frictional forces create agglomerated products. Commonly low and medium pressure agglomerations are carried out by different extruders. The screen, basket, and cylindrical die screw extruders are applied at low pressure conditions, while at medium pressure agglomeration flat–die, cylindrical die, and intermeshing gears extruders are used.
- High pressure agglomeration is characterized by a large degree of densification, low-porosity agglomerates, high strength of agglomerates and pillow or almond–like shape products. The use of post–treatment methods or little amounts of binders can further increase the agglomerate's strength. High pressure agglomeration is a successful method to agglomerate of any type and size of powders (from

Table 7.2 The advantages and disadvantages of pressure agglomeration [Adapted from Saravacos and Kostaropoulos 2002]

Advantages	Disadvantages
Raising of agglomerates strength	High energy demand
Diversity of shape and size	Low processing capacity
Increase the product density	Wear-out of equipment and tools
Use of powders that cannot be agglomerated otherwise	High cost of auxiliary tools(dies, molds)

nanometers to centimeters). High pressure agglomeration is carried out by pressing including the compacting roller press, the briquetting roller press and the punch–and–die press.

7.3.1.1 Extrusion Techniques

The extrusion agglomeration forces a powder mixture (powder, binder liquids, additives, or dispersants) to flow through a die (shaped hole) at the specific rate at low pressure followed by drying, cooling, and crumbling operations in which raw materials undergo definite shear and thermal energy as it is consolidated while being compact through a die (Fig. 7.3). The structural, chemical and nutritional properties of raw materials change during the extrusion like starch gelatinization and aroma formation. The extrusion agglomeration can produce a large number of agglomerated foods with different size, shape, texture, and taste such as pasta, cereal–based foods, bread crumbs, baby foods, biscuits, crackers, snack foods, chewing gum, modified starch, dried soups, encapsulated vitamins and flavors, pet foods, and dry beverage mixes. Also, extrusion modifies the water solubility, swelling properties, water holding capacity, water absorption, oil absorption indexes, and moisture hydration properties of agglomerated products (Alam et al. 2015).

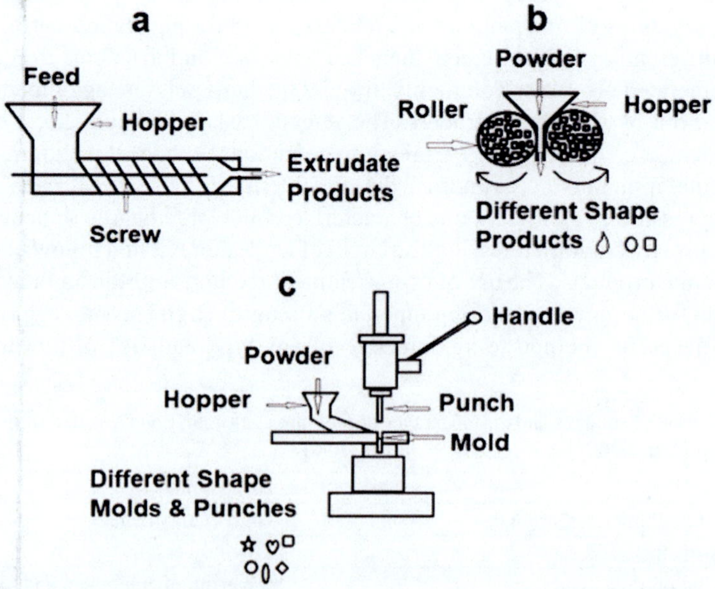

Fig. 7.3 Equipment for pressure agglomeration: (**a**) Extruder, (**b**) Structured Roll Press, c Manual Tableting Press [adapted from Barbosa-Canovas et al. 2005]

7.3.1.2 Roll Pressing Methods

Roll pressing agglomeration equipment presses raw powder as it is passed through the gap between two reversely rotating rolls which rotates at the same speed. The main advantage of roll press machines is the usage of the low amount of material. The size and shape of the agglomerated particles are depended on the geometry of the roll surfaces. In structured pressing rolls, pockets or indentations in the surfaces of cylinders (rolls) produce eggs, pillows or teardrops briquettes (Fig. 7.3) (Green 2007). While smooth or corrugated surfaces create a solid sheet which later crushed into the desired size by grinding machine. The high speed rotation increases the released air which produces fluidized particles and reduces the homogeneity of the agglomerates. Therefore, generally, the rotation speed maintains between 5 and 40 rpm. The necessary pressure in the smooth rollers for compressing dry powder is between 1 and 14 kbar, while for mist powder is varied from 1 to 100 bars. The roll pressing briquettes can affect from different factors such as treatment temperature, raw material properties (size, shape, size distribution and surface), type of equipment, binders, and properties of agglomerated products (moisture content, hardness and brittleness) (Saravacos and Kostaropoulos 2002).

7.3.1.3 Tableting Press

Tableting press are used in the production of materials with strict specifications for weight, density, thickness, strength, and forms. The tableting machine consists of filling funnel, upper punch, lower punch, and molds. The powder is poured from filling funnel into the molds and compressed between two pistons (Fig. 7.3). The funnel can be fixed while the pistons and molds are movements or the funnel moving from one mold to others while the pistons and molds are fixed. However, feed material needs to have high flowability to uniformly fill the molds before the compression process. Variables influencing tableting presses are flow attributes of feed, type and quantity of binders, adhesiveness of formed tablets to the pistons, and ease of tablet removal at the end of the process (Saravacos and Kostaropoulos 2002; Dhanalakshmi et al. 2011).

7.3.1.4 Pelletizing Methods

Pelletizing agglomeration units combine the powder particles by pressing and pushing them through an opening in dies with different shapes. The rod-like pellets are cut to the desired length by an adjustable knife (Fig. 7.4). However, diverse designs are in use, the two most basic pelletizing techniques are as follows:

1. Rotating blade or rolls pushed the material through a fixed screen. The type of raw and agglomerated materials are the factors for determination of machine capacity. Pellet diameter is ranged from 1 mm to larger than 5 cm.

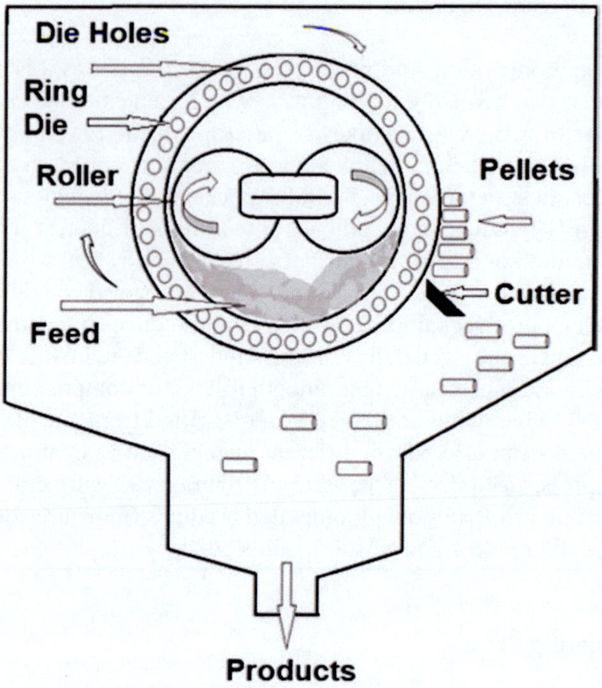

Fig. 7.4 Pelletizing Equipment [adapted from Green 2007]

2. The feed is pressed among perforated hole cylinders which are often used to agglomerate the moist materials. The counter rotating of a non-perforated cylinder pressed the feeds among the perforated cylinders. The formed pellet diameter is usually larger than 5 cm.

Pellet quality and capacity are affected from characteristics of the feeds (particle size, moisture, abrasiveness), resistance against passing the material through holes, the residence time of feed in the holes, binders, size of agglomerates, applied pressure, die characteristics (Saravacos and Kostaropoulos 2002; Green 2007).

7.3.1.5 Instant Agglomerates by Press Methods

In press agglomeration methods two techniques can apply to produce instant agglomerates:

1. Compaction/granulation technique: Firstly, a mixture of dry particles is compacted by high pressure. After that, the compressed agglomerates crushed and screened into the granule with the desired size and other operations are not necessary. These products have a high density.

2. Extrusion/crumbling technique is used when compaction/granulation methods are not possible. At this method mixture of dry particles first mix with a liquid binder and then extruded at low pressure follow drying, cooling and crumbling.

In these techniques, the main mechanisms to bond particles together are intermolecular (van der Waals) or intramolecular (electrostatic) forces rather than solid bridges. When these agglomerates dissolved in liquids, these short range molecular forces are decreased by a factor of around ten times and the agglomerates can fastly disperse in a liquid medium exhibiting instant characteristics of agglomerates (Dhanalakshmi et al. 2011).

7.3.2 Tumbling and Mixing Agglomeration (Rewetting Agglomeration)

Tumbling, vibrating, shaking or mixing of particles in the presence of a sufficient amount of liquid (binders/solvents) formed a wet mass of particles by adhesion. Wet granulation process consists of different steps: (1) wetting and nucleation step, (2) growth and coalescence, (3) consolidation and breakage phases, (4) drying stage (Thapa et al. 2019). In the first step, dry particles are wetted by the liquid binder and adhesion force, stick wetted powder and formed small aggregates so–called nucleus. At the growth and consolidation stage, the particles collide in equipment which increases the size and volume of the aggregates by either coalescence, layering or with both of them (De Simone et al. 2018). Layering means that the layers of raw material deposits on the formed nucleus surfaces. At the consolidation stage, the compaction forces (due to pressure or agitation) consolidate the granules which influence porosity, dispersibility and strength of granule. The products may be susceptible to breakage due to disintegrating of the weaker bonds and re-agglomerate by coalescence and abrasion transfer. However, the simultaneous interaction of all these processes gives complexity to this operation. Finally, the drying stage separates the wetting phase from granules and increases the strength and stability of the products. However, the formation of a stable nucleus from fine powder is difficult due to little coordination points in small granules. Furthermore, the kinetic energy of powder and small nuclei is not sufficient to improve bonding at contact points. Therefore, the undersized fines can be recirculated to produce suitable nuclei for easy adhesion of feed to them (Barbosa-Canovas et al. 2005).

The most common types of equipment used are rotating drums, discs, pans, and any powder mixers. In the rolling agglomeration, rolling up of the powder is the basis of the process. The rotating walls cause the feeds to roll up, fall down and take upward motion again. The vibrating inclined bed of some equipment also can give the rolling motion to granules as well. However, in the mixing techniques, contact with raw materials is the main reason for their enlargements. Factors that influence the tumbling and mixing agglomerations are: the type and amount of binders, the size of binder droplets and dry powders, the temperature of liquid binders and dry particles, and the position of binder and powder are added (Saravacos and Kostaropoulos 2002). Although, in all the above processes, surface tension and capillary forces make a temporary bond therefore, weak agglomerates (green agglomerates) form. Some post–treatment can be used to increase the strength of these agglomerates by drying, heating, cooling, screening, improve the properties of the product by crushing, rescreening, and recirculation of undersize products (Barbosa-Canovas et al. 2005).

7.3.2.1 Inclined Pan or Disc Equipment

The inclined pan or disc rotates around their inclined axis. Dry powder and binder are fed at the upper section of the equipment. The liquid binders continuously sprayed on the products. To ensure correct tumbling operation, the inner surface of the pan or disc must be rough (Fig. 7.5). The agglomerates leaving the equipment are spherical with the same size with a diameter between 0.5 and 2 mm. For most applications, the pan angle preferred to adjust at a range of 45–55° (0.78–0.96 rad). The maximum diameter of the disc in the food industry is 2–3 m. The efficiency of pan agglomeration is affected by pan angle, rotation speed, location of addition dry powder and liquid binder, and the *D/h* ratio (*D* and *h* are the diameters and the edge of the pan). The *D/h* ratio generally ranged from 0.08 to 0.5 (Saravacos and Kostaropoulos 2002).

7.3.2.2 Drum Machines

The rotating drum agglomerator works like a pan agglomerator but the homogeneity of the product is different. The size of granules is not the same and need to screen out the products and send the fine agglomerate to the rotating drum to achieve the desired size. At this method, an inclined cylinder rotates at an angle up to about 0.175 rad (10°) (Fig. 7.5). It is possible to moistened feed in various mixers to form ball nuclei and then pass them to the drum, or liquid binder may be directly sprayed

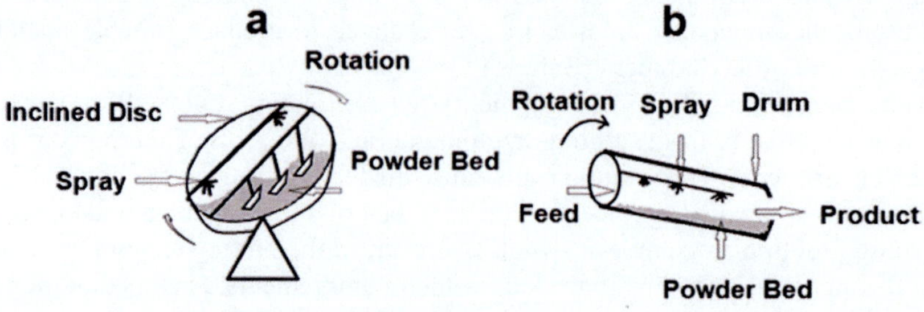

Fig. 7.5 Tumble/growth agglomeration equipment: (**a**) Inclined disc, (**b**) Drum agglomerator [adapted from Saravacos and Kostaropoulos 2002]

on the rolled product in drum agglomerator. Commonly, the drum (*L/D*) ratio is 2–5 (L = drum length, *D* = drum diameter) (Saravacos and Kostaropoulos 2002).

7.3.2.3 Mixer Equipment

Agglomerates can form by shear processing in many types of horizontal or vertical mixers. Like as planetary mixers, ribbon blenders, Z-blade units, agitating systems, high–speed intensive mixers, pug mixer, angular paddles, and ploughs (Dhanalakshmi et al. 2011). The size of agglomerates is affected by the amount of liquid binders, mixing intensity and mixing time. Intensive mixers produce a rubbing and shearing effect on the powder which helps to produce more strength agglomerates when compared with tumbling method. Products in the rotating mixers are mechanically fluidized which decreases process time.

The paddles, bars and rods are attached to one or more shafts inside the paddle mixer. For example, a twin shaft paddle agglomerator contains two shafts rotate in the opposite direction in a vat. Agglomeration can be complete after 1 min. These equipment are not depended on the size and density of agglomerates (Green 2007; Saravacos and Kostaropoulos 2002).

The plough mixer consists of a cylindrical vat and a single shaft with a rotating speed of 60–800 rpm. Also, there are some choppers that attached to the cylinder walls to reduce the lump formation during the process (especially for a product with high moisture content). These choppers rotate vertically with a speed of more than 3000 rpm. Agglomeration is finished at 1–4 min. They had wide applications and can produce agglomerates from 1 to 1000 micrometers.

The Ruberg mixer (two helices) consist of two adjoined cylinder with two large helical agitators. The feed with help of the first shaft goes to the down part of the equipment then spirally moves upward around the wall of the first cylinder and subsequently moves along the second shaft downward and again moves upward around the wall of second vats. Feeds move in counterflow among two cylinders and agglomeration lasts up to 3 min. The gentle mixing and simple construction are the main advantages of these agglomerators (Saravacos and Kostaropoulos 2002).

Low speed agitators belong to batch type operation and they are conical planetary mixer which designed as a double-wall mixer to control the product temperature. The slowly moving mixer rolled the feed materials and agglomeration is finished at 20–40 min. It is a batch type operation.

Vibration agglomeration is a continuous process in which the feeds distribute on the vibrating bed equipped with some barriers. Small agglomerates leap over barriers by vibration and a few times they roll back before to go to the next barrier and being enlarged and leave the vibrating bed.

7.3.2.4 Instant Agglomerates by Non-pressurized Techniques

The main aim of instantizing is to modify the rate and completeness of the reconstitutability of dried agglomerates. In many cases, the instantizing processes were preferred by kept the fine powder in a fluidized state by the agitation methods such as mechanically agitation (mixers) or agitation by gas (fluidized bed). The rewetting agglomeration techniques can form products with great re–dispersion properties.

The common steps are:

- Rewetting of the powder surface with steam, atomized liquid binders, or a combination of both.
- Collide the particles due to turbulence and formed liquid and solid bridges which produce clusters
- Drying with hot air (through the mixer or fluidization unit)
- Cooling and screening

In the food industry, water is used generally as a binder but, for water insoluble powders special binders can be used. Besides water, sugar, starch, molasses, gelatin, Arabic gum, and maltodextrin solutions are used as food binders. To improve the wetting process, the binder droplet size must be equal or smaller than the powder particles.

Also surface techniques can be preferred for rewetting methods in which powder from a vibratory feeder, poured on the rotating disk and as powder fall down from disc, it moistened by injection of moist steam or spraying a binder into the powder stream. For products with high hygroscopic natures (coffee, coffee substitute) the disk is rotated at high speed (500–1200 rpm) while for products with high hygroscopic properties its speed is (20–50 rpm) (Barbosa-Canovas et al. 2005).

7.3.3 *Spray Agglomeration*

Spray methods can be used to agglomerate fine powder particles. Fundamentally, solid feed in a fluid state (solution, slurry, emulsion) is dispersed in a gas flow and converted to the agglomerated products by heat and/or mass transfer. The maximum agglomerate diameter is 5 mm. Spray drying and fluidized bed granulation are the most common techniques used in the industry for agglomeration.

7.3.3.1 Spray Drying

The feed is suspended in the air at the dryer chamber and dried in a very short time. This process takes place in three steps as (1) atomization step, (2) mixing and contacting step (feed and air), (3) separation of product from the gas stream. Atomization can be approached by rotary wheel (centrifugal disc), pressure nozzle (single fluid), or pneumatic nozzles (two fluid). During the process, moisture evaporates from the feed surface and viscosity increases progressively and at some critical value, surface of particle shows sticky state in which collision of sticky particle together produces agglomerate products. Surface stickiness is affected by temperature, moisture content and feed composition (carbohydrates). Furthermore, the angle and time of contact between particles, forces and contact velocity all have an impact on the formation of agglomerates (Gianfrancesco et al. 2008). The agglomerates size is a function of atomization condition (disc speed, nozzles size), solid content, feed rate, density and viscosity of feed. Increasing viscosity and feed rate and the presence of binder can increase product size (Green 2007).

7.3.3.2 Fluidized Bed Granulation

In this method, the powder particles are fluidized by the preheated air which is blown from the bottom of the equipment and a binder solution is sprayed onto the dispersed particles from the top, bottom, and side of the chamber. The top-spray, bottom-spray, tangential-spray, Wurster–coater (Recycling fluidized bed agglomerators), and spouted bed are the most common used agglomeration equipment (Thapa et al. 2019). Two mechanisms may be involved in particle-growth as (1) layering of solid on single particles, (2) agglomeration of several particles to form a larger particle. Due to multiple deposited layers on the particle, the size of agglomerated products is larger than that produced by spray dryer (Green 2007). Agglomeration by fluidized bed drying is affected by the rate of binder, binder droplet size, air velocity, bed high and temperature. Fluidized is achieved when operating air velocity (u) is in the range of 1.5 $u_m < u < u_p$, where u_m is the minimum air velocity and u_p is the air velocity at which particles are pneumatically conveyed. The necessary air velocity u (m/s) to fluidize a bed of particles can be calculated by the Eq. 7.7 (Saravacos and Kostaropoulos 2002).

$$u = \left[\frac{(\rho_s - \rho) g}{\eta} \right] \left[\left(d^2 \epsilon^3 \right) \left(180 \left(1 - \epsilon \right) \right) \right] \tag{7.7}$$

where ρ_s, ρ, g, η, d, and ϵ are the density of particles (kg/m^3), the density of air (kg/m^3), the gravitational acceleration (m/s^2), the fluid viscosity (Pa s), the particle diameter (m), the bed porosity, respectively.

7.3.4 *Alternative Agglomeration Methods*

7.3.4.1 Steam Granulation

In steam jet agglomeration powder particles are wetted with water steam and are used instead of traditional liquid binders. The viscous layers on the surface of wetted powder during the colliding in the wetting zone formed agglomerates. Furthermore, if the dry feed particles bonded by van der Waals forces and formed agglomerates before enter to the steam jet equipment, these dry agglomerates also take up water and become more strength as well as dry fine powder particles (Dhanalakshmi et al. 2011). Also, as steam condensed, a hot thin water layer formed on the particles, which with extra drying can easily eliminate. The drying step crystallized the soluble substances in the liquid bridges and produces solid bridges. This method can produce spherical granules with a high surface area which leads to a higher dissolution rate (instant properties) compared to conventional wet agglomeration methods.

7.3.4.2 Thermal Adhesion Agglomeration

In this technique mixture of dry powder and a small amount of granulation liquid (binders, water) is heated (30–130 °C) under the tumble rotation to form the wet agglomerated products and after that cooling, the products produce the dry agglomerates (Shanmugam 2015). If the amount of used liquid is low, the expensive drying step is not necessary. This technique is simple and can improve flow properties, tensile strength and hardness of products.

7.3.4.3 Freeze Agglomeration

In this method, a powder suspension is sprayed into liquid nitrogen and droplets are immediately frozen into granules. At the next step, frozen granules are dried in the freeze dryer. Sublimation of ice from granules creates products with high porosity, high homogeneity and very good instant properties such as instant coffee, tea extract and fruit juices.

7.4 Criteria for Selection of Agglomeration Processes

The main criteria to select suitable agglomeration processes, optimizing operations, improving the functionality and decreasing product cost are properties of feed, and agglomerated products as well as the methods. Parameters that determine the properties of raw material are powder properties (particle size and size distribution, density, shape, porosity, surface area, roughness), bulk properties (bulk density, flow properties), textural attributes (brittleness), rheological properties (flowability, elasticity, plasticity, surface tension), water content of powder, wettability, physical and chemical properties, surface characteristics, geometry, and etc. (Barbosa-Canovas et al. 2005). The main product variables are agglomerates shape, size and size distribution, porosity, density (bulk & particle) flowability, and strength of the agglomerates. Technical criteria are batch/continuous process, capacity, wet/dry methods, applied force, mixing, drying, fluidization, shear or impact rate, temperature and residence time. Also, economic criteria must be considered for the selection of agglomeration methods and equipment such as the cost of equipment, energy consumption, cost of binders (Saravacos and Kostaropoulos 2002).

7.4.1 Raw Material Characteristics

7.4.1.1 Size and Size Distribution

For considering the particulate material as a powder, its median size should be less than 1 mm. The particle size and size distribution have an important effect on the many properties of powder such as bulk density, flowability, and compressibility. The general method for particle size measurement is sieving, microscope counting techniques, sedimentation, and stream scanning (Dhanalakshmi et al. 2011). Generally, the growth agglomeration based on coalescence is more suitable for fine powders with the size of a few hundred micrometers. However, for larger powders use of a sufficient amount of binder or existence of enough amount of small particles are necessary to produce seed agglomerates. In this case, the strength of agglomerate is related to the matrix produced by embed fine powders in larger particles. For the particle up to 20–30 mm, pressure agglomeration is preferred (Barbosa-Canovas et al. 2005). Size distribution has an important effect on the process as an agglomeration of particles with mono-sized or narrow–sized distribution is tough. Furthermore, the crush of larger particles for the growth agglomeration process help to produce the product with higher porosity.

7.4.1.2 Moisture Content

- In the growth agglomeration, free water binds particle together by coalescence which the binder volume is restricted to less than 95% of anticipated agglomerate porosity.
- Wet agglomeration has high sensitivity to moisture content and moisture excess than critical level gives mud-like consistency to whole products.
- In fluidized bed agglomeration, the moisture is only necessary for providing pumpable property to raw material.
- In pressure agglomeration, moisture must be low because the water is incompressible component and compression forces only cause crushing and deformation of solid raw material. Therefore, excess water remains in the compacted products and decreases the agglomeration strength.

7.4.1.3 Material Properties

- Knowledge about chemical properties of powder is important in agglomeration. According to the chemical properties, a particle can bind together with chemical bonding or addition of water or other liquids that may be avoided in growth agglomeration due to the characteristics of particles.
- The density of the particle is another property that must consider for agglomeration. It can determine the field forces that may be counteractive to addition by coalescence.
- Textural (brittleness) and rheological (elasticity, plasticity) properties of feed are important criteria in choosing pressure agglomeration.
- However, the special notice must be done on wettability for agglomeration techniques in which surface tension and capillary forces are an important parameter in the formation of agglomerates.

7.4.2 Characteristics of Agglomerated Products

- One of the basic parameters for the evaluation of the quality of agglomerated products and the performance of the operation is particle size. Among different methods for determination of size, sieving is often preferred due to its simplicity. Approximately, the size of many agglomerates is around 1 mm (Barbosa-Canovas et al. 2005). After the agglomeration process, narrowing the size distribution of products provide by screening to separate under and oversized agglomerates. Undersized products are recirculated to the agglomerator, while oversized products are crushed to the desired size.
- Generally, the demand for spherical agglomerate is more than other shapes. The spherical products can be successfully formed by growth agglomeration methods while high pressure agglomeration cannot create spherical agglomerates. However, approximately spherical shape such as a pillow, lens, or almond-shaped compacts can produce by tableting equipment.
- The flowing properties of agglomerated products are affected from surface properties, shape, size distribution and forces (gravity, adhesion, cohesion) (Dhanalakshmi et al. 2011). Free–flowing or dust–free agglomerates can be produced using almost all agglomeration techniques.

7.4.3 Characteristics of Methods

Batch or continuous agglomeration process may be preferred according to special requirements. Generally, the capacity of the batch process is lower than continuous processes but control of the operation in batch type equipment is better than continuous ones. Also, agglomeration methods can be selected based on wet processes (growth agglomeration) or dry operation (high–pressure agglomeration). Besides, other processes may be carried out simultaneously with agglomeration like mixing, drying and fluidizing or combination of them. However, shear rate, temperature, residence time and impact rate are the other technical criteria that need to be considered for the selection of agglomeration techniques.

7.5 Instantizing Processes

In the food industry, the term "instant" is usually described as the dispersing and dissolving attributes of food powders such as milk, tea, soups, coffee, sauces, starches, sugars, dextrins, dry pudding mixes, cocoa, baby foods, juices, etc. Instantizing process will produce products with the "instant" property which means food powder can be dissolved faster and more readily than in their original powdered forms (Barbosa-Canovas et al. 2005). Instantizing process can be divided into two groups as agglomeration and non-agglomeration.

7.5.1 Agglomeration Processes

Different agglomeration processes were discussed previously. Product with instant properties can be obtained by pressure agglomeration, re–wetting agglomeration, and spray drying mehods.

7.5.2 Non–Agglomeration Processes

Non–agglomeration processes for production of instant powders such as freeze drying, thermal treatment of amorphous foods, osmotic and drum drying, separation of fat from products, or adding some additives (lecithin) (Barbosa-Canovas et al. 2005).

- *Freeze-drying*: Sublimation of ice crystals from products during the freeze drying produce particles with high porosity and an open pore structure which improves ease of absorbing water.
- *Osmotic drying*: In this method, food particles are immersed in the osmotic solution and water inside the particle is transferred to the solution because of the osmotic pressure. After that, these partially dehydrated products are exposed to final drying. This combined method produces the products with an open pore structure which will display good reconstitution properties.

7.5.3 *Instant Properties*

Among the functional properties of food powder, rehydration and reconstitution have special importance. When instantized powder particles are spread on the surface of a liquid or water, reconstitution or dissolution properties it can be determined according to the four steps such as wettability, sinkability, dispersibility, and solubility.

7.5.3.1 Wettability

Wettability is the first step in the reconstitution of food powders and it is the rate–controlling stage. It refers to wetting the particles by absorbing a liquid on their surface due to capillary suction of the pores in products, therefore, initiating reconstitution. The liquid penetrates into the particles due to the ability of the particle to overcome the surface tension between powder particles and liquid (Fang et al. 2007). The wetting property mainly depends on the (1) particle size, and (2) nature of the particle surface. Large surface area of small particles causes that powder particles did not wet individually thus they produce undesired clump structure. Also, when free fat exists on the particle surface wettability is reduced. However, the use of some active- agents (lecithin) in fatty powder can improve the wetting property (Barbosa-Canovas et al. 2005). The desired conditions for good wettability are:

- Large particle
- Large pores between particles
- High porosity (bellow a critical porosity)
- Small contact angel

7.5.3.2 Sinkability

Sinkability means that after the powder has been wetted they fell below the liquid surface. At this stage, the air surrounding each particle is substituted by the liquid phase and causes the particle sink quickly into the liquid. Sinkability is increased by increasing the size and density of particles. A higher particle density along with the presence of a smaller amount of air enclosed within the powder particle can increase the sinking rate (Fang et al. 2007).

7.5.3.3 Dispersibility

After the powder particles are wetted and have sunk, they would start to distribute as single particles over the liquid surface which refer to dispersibility (Barbosa-Canovas et al. 2005). Increasing the particle size and sinkability and reducing the percentage of fine particles increase the dispersibility. Particles with high porosity and density improve the dispersibility property of products. However, clump formation can decrease dispersibility (Fang et al. 2007).

7.5.3.4 Solubility

Solubility is the terminal stage in the reconstitution of food powders and refers to the dissolving of soluble powder in the liquids which is generally related to the chemical composition and physical state of the powder. As mentioned before, dispersibility refers to the ease of particle distribution while solubility is indicated the rate and extent of dissolution (Dhanalakshmi et al. 2011).

7.6 Porous Powders

Porous powders are developed in recent years due to their important characteristics like large surface area, high porosity, stable homogenous porous structure, sizeable pores and adequate surface properties (Zhou et al. 2017). Due to their multifaceted characteristics, they are used as an effective carrier agents in the pharmaceutical and food applications. They vary depending on pore diameter sizes and classified as (1) macroporous (diameter > 50 nm), (2) mesoporous (2 < diameter < 50 nm) and (3) microporous (diameter < 2 nm). Production methods and application of porous particles are well summarised in the review paper published by Zhou et al. (2017). There are some studies about production of porous powders and their applications: Porus silicon (Salonen et al. 2005), porous silica (Tanimura et al. 2015; Yan et al. 2015), porous mannitol (Saffari et al. 2016; Leung et al. 2017), porous lactose (Ebrahimi et al. 2015, 2017), porous starch (Belingheri et al. 2012; Zhang et al. 2012), porous chitosan (Li et al. 2017). Starch is an widely used agent in industrial applications especially in food industry. It can be used as a thickener, stabilizer, gelling agent, etc. Generally, these type of starches are called as native starches and they produce cohesive, weak-bodied, rubbery pastes with heating and after the cooling (Dura et al. 2014). Thanks to their better adsorption ability than native and modified starches, porous starches are very popular in these days among the industries such as food, medicine, agricultura, cosmetics etc. (Zhang et al. 2012). These starches are used as spices, protection of sensitive substances (e.g. oils, pigments, minerals, vitamins, bioactive lipids) and carrier agents for sweeteners and flavours in the food industry (Benavent-Gil and Rosell 2017). These starches can be also used for encapsulation purposes of some biomaterials for protection from the environmental conditions and controlled release (Glenn et al. 2010). By increasing the specific surface area with abundant pores in the starch, it is possible to have excellent natural absorbents (Benavent-Gil and Rosell 2017). Porous starches can be produced using physical, chemical and biological methods. Among them, ultrasonic and enzyme treatment are reported as the mostly used methods.

An enzyme based method is the most common method for producing microporous starch (Wu et al. 2011). The advantages of enzymatic modifications are high yield, fewer by–products, more specificity and better control over the process. For altering starch structure and achieve desired functionality number of enzymes can be used such as α–amylase, β–amylase, glucoamylase, pullulanase, and isoamylase (Dura et al. 2014). Benavent-Gil and Rosell (2017) stated that the properties of porous starches such as pore size distribution and pore area could be modified by using different enzymes and concentrations. In addition, starches can be modified by replacing ice crystals in frozen starch gel with ethanol/water mixture used as solvent. The hole size of the porous starch can be altered with the change in ethanol/water ratios. Additionally, by adjusting the number of freezing cycles and starch paste concentrations, porosity of porous starch can be changed as desired (Chang et al. 2012). As there are some applications of porous starches in the encapsulation of sensitive functional ingredients and flavor compounds, the performance of it into real food systems. And these studies should be also proved by bioaccessibility and bioavailability analyses to better understand the advantages and disadvantages of porous starch.

Porous lactose is also studied in limited studies. Ebrahimi et al. (2015) produced highly-porous lactose using sugars as templating agents. And in their other study, porous lactose is used as an improving dissolution rate of hydrophobic drugs as a new biocompatible porous carrier (Ebrahimi et al. 2017). This study can be applied to the food industry for encapsulation of hydrophobic functional compounds for improving dissolution rate of them.

7.7 Conclusion

Food powders are widely used in the food industry as both intermediate and final products. Their solubility, flowability, surface, density, compressibility, hydration and particle size characteristics determine their usage area and amount in the formulation. These physicochemical properties can be improved by modification of the powders. Enlargement operations specially agglomeration have much application in food powder industries. They used to control powder properties (porosity, heat transfer rates, solubility), decrease dusting or product losses and improve powder transport and storage. Also, agglomeration can significantly improve flow properties, dispersion and dissolution properties of the powder and attained to the uniformity in dosage and composition of products. Properties of powder particles, agglomerates and techniques directly affect the equipment selection, storage ability and formulation. The reconstitutability of many products in the food industry such as instant coffee, fruit drinks, starch, dried milk, dried soups and pudding, sugar, cocoa powder, tea, flour and etc., can be improved by agglomeration processes. Agglomeration with extrusion methods has been extensively applied in the grain industry to produces the ready-to-eat breakfast cereals with a broad range of shapes. Taking into account the concepts presented above, it is obvious that the agglomeration has a great influence on the modification of powder particles in the food, pharmaceuticals, detergents, agricultural, nutraceuticals, cosmetics, and mineral industry. The increased surface area of the powders such as starch and lactose can provide an opportunity to use them in encapsulation of bioactive compounds and flavor compounds for preventing them against stress of environmental conditions such as temperature, pH and light and controlling release or dissolution of these compounds. Usage of such porous powders in the food industry is a pretty new subject which should be studied deeply in real food systems.

References

Aguilar-Palazuelos, E., Zazueta-Morales, J. D. J., Harumi, E. N., & Martínez-Bustos, F. (2012). Optimization of extrusion process for production of nutritious pellets. *Food Science and Technology, 32*(1), 34–42. https://doi.org/10.1590/s0101-20612012005000005.

Alam, M. S., Kaur, J., Khaira, H., & Gupta, K. (2015). Extrusion and extruded products: Changes in quality attributes as affected by extrusion process parameters: A review. *Critical Reviews in Food Science and Nutrition, 56*(3), 445–473. https://doi.org/10.1080/10408398.2013.779568.

Augsburger, L. L., & Hoag, S. W. (2008). *Pharmaceutical dosage forms - tablets*. Boca Raton: CRC Press. https://doi.org/10.3109/9781420020304.

Barbosa-Canovas, G. V., Ortega-Rivas, E., Juliano, P., & Yan, H. (2005). *Food powders: Physical properties, processing, and functionality, Food Engineering Series*. New York: Kluwer Academic/Plenum Publishers.

Belingheri, C., Curti, E., Ferrillo, A., & Vittadini, E. (2012). Evaluation of porous starch as a flavour carrier. *Food & Function, 3*(3), 255–261. https://doi.org/10.1039/c1fo10184f.

Benavent-Gil, Y., & Rosell, C. M. (2017). Comparison of porous starches obtained from different enzyme types and levels. *Carbohydrate Polymers, 157*, 533–540. https://doi.org/10.1016/j.carbpol.2016.10.047.

Buffo, R. A., Probst, K., Zehentbauer, G., Luo, Z., & Reineccius, G. A. (2002). Effects of agglomeration on the properties of spray-dried encapsulated flavours. *Flavour and Fragrance Journal, 17*(4), 292–299. https://doi.org/10.1002/ffj.1098.

Chang, P. R., Qian, D., Anderson, D. P. and Ma, X. (2012) Preparation and properties of the succinic ester of porous starch, Carbohydrate Polymer, 88(2), pp. 604–608, https://doi.org/10.1016/j.carbpol.2012.01.001

De Simone, V., Caccavo, D., Dalmoro, A., Lamberti, G., d'Amore, M., & Barba, A. A. (2018). Inside the phenomenological aspects of wet granulation: Role of process parameters. *Granularity in Materials Science*. https://doi.org/10.5772/intechopen.79840.

Dhanalakshmi, K., Ghosal, S., & Bhattacharya, S. (2011). Agglomeration of food powder and applications. *Critical Reviews in Food Science and Nutrition, 432*(441), 51.

Dura, A., Błaszczak, W., & Rosell, C. M. (2014). Functionality of porous starch obtained by amylase or amyloglucosidase treatments. *Carbohydrate Polymers, 101*, 837–845. https://doi.org/10.1016/j.carbpol.2013.10.013.

Ebrahimi, A., Saffari, M., & Langrish, T. (2015). Spray drying and post-processing production of highly-porous lactose particles using sugars as templating agents. *Powder Technology, 283*, 171–177. https://doi.org/10.1016/j.powtec.2015.05.026.

Ebrahimi, A., Saffari, M., & Langrish, T. (2017). Improving the dissolution rate of hydrophobic drugs through encapsulation in porous lactose as a new biocompatible porous carrier. *International Journal of Pharmaceutics, 521*(1–2), 204–213. https://doi.org/10.1016/j.ijpharm.2017.02.052.

Ennis, B. J. (1996). Agglomeration and size enlargement session summary paper. *Powder Technology, 88*(3), 203–225. https://doi.org/10.1016/s0032-5910(96)03124-5.

Fang, Y., Selomulya, C., & Chen, X. D. (2007). On measurement of food powder reconstitution properties. *Drying Technology, 26*(1), 3–14. https://doi.org/10.1080/07373930701780928.

Feng, J. Q., & Hays, D. A. (2003). Relative importance of electrostatic forces on powder particles. *Powder Technology, 135–136*, 65–75. https://doi.org/10.1016/j.powtec.2003.08.005.

Gianfrancesco, A., Turchiuli, C., & Dumoulin, E. (2008). Powder agglomeration during the spray-drying process: Measurements of air properties. *Dairy Science & Technology, 88*(1), 53–64. https://doi.org/10.1051/dst:2007008.

Glenn, G. M., Klamczynski, A. P., Woods, D. F., Chiou, B., Orts, W. J., & Imam, S. H. (2010). Encapsulation of plant oils in porous starch microspheres. *Journal of Agricultural and Food Chemistry, 58*(7), 4180–4184. https://doi.org/10.1021/jf9037826.

Green, D. W. (2007). *Perry's chemical engineers' handbook*. New York: McGraw-Hill.

Knight, P. C. (2001). Structuring agglomerated products for improved performance. *Powder Technology, 119*(1), 14–25. https://doi.org/10.1016/s0032-5910(01)00400-4.

Kundu, K., Chatterjee, A., Bhattacharyya, T., Roy, M., & Kaur, A. (2017). Thermochemical conversion of biomass to bioenergy: A review. *Prospects of Alternative Transportation Fuels*, 235–268. https://doi.org/10.1007/978-981-10-7518-6_11.

Leung, S. S. Y., Wong, J., Guerra, H. V., Samnick, K., Prud'homme, R. K., & Chan, H.-K. (2017). Porous mannitol carrier for pulmonary delivery of cyclosporine a nanoparticles. *The AAPS Journal, 19*(2), 578–586. https://doi.org/10.1208/s12248-016-0039-3.

Li, J., Wu, X., Wu, Y., Tang, Z., Sun, X., Pan, M., Chen, Y., Li, J., Xiao, R., Wang, Z., & Liu, H. (2017). Porous chitosan microspheres for application as quick in vitro and in vivo hemostat. *Materials Science and Engineering: C, 77*, 411–419. https://doi.org/10.1016/j.msec.2017.03.276.

Popescu, I. N., & Vidu, R. (2018). Densification mechanism, elastic-plastic deformations and stress-strain relations of compacted metal-ceramic powder mixtures (review). *Scientific Bulletin of Valahia University-Materials and Mechanics, 16*(14), 7–12. https://doi.org/10.1515/bsmm-2018-0001.

Reid, R. C. (1974). In R. H. Perry, C. H. Chilton, & S. D. Kirkpatrick (Eds.), *Chemical engineers' handbook* (2nd ed.). New York: McGraw-Hill. https://doi.org/10.1002/aic.690200140.

Saffari, M., Ebrahimi, A., & Langrish, T. (2016). A novel formulation for solubility and content uniformity enhancement of poorly water-soluble drugs using highly-porous mannitol. *European Journal of Pharmaceutical Sciences, 83*, 52–61. https://doi.org/10.1016/j.ejps.2015.12.016.

Salonen, J., Laitinen, L., Kaukonen, A. M., Tuura, J., Björkqvist, M., Heikkilä, T., Vähä-Heikkilä, K., Hirvonen, J., & Lehto, V.-P. (2005). Mesoporous silicon microparticles for oral drug delivery: Loading and release of five model drugs. *Journal of Controlled Release, 108*(2–3), 362–374. https://doi.org/10.1016/j.jconrel.2005.08.017.

Saravacos, G. D., & Kostaropoulos, A. E. (2002). Handbook of food processing equipment. *Food Engineering Series*. https://doi.org/10.1007/978-1-4615-0725-3.

Shanmugam, S. (2015). Granulation techniques and technologies: Recent progresses. *BioImpacts: BI, 5*(1), 55–63. https://pubmed.ncbi.nlm.nih.gov/25901297.

Simons, S. J. R. (2007). Liquid bridges in granules, in granulation. In *Handbook of powder technology* (Vol. 11). Amsterdam: Elsevier.

Tanimura, S., Tahara, K., & Takeuchi, H. (2015). Spray-dried composite particles of erythritol and porous silica for orally disintegrating tablets prepared by direct tableting. *Powder Technology, 286*, 444–450. https://doi.org/10.1016/j.powtec.2015.08.011.

Tardos, G. I., Khan, M. I., & Mort, P. R. (1997). Critical parameters and limiting conditions in binder granulation of fine powders. *Powder Technology, 94*(3), 245–258. https://doi.org/10.1016/s0032-5910(97)03321-4.

Thapa, P., Tripathi, J., & Jeong, S. H. (2019). Recent trends and future perspective of pharmaceutical wet granulation for better process understanding and product development. *Powder Technology, 344*, 864–882. https://doi.org/10.1016/j.powtec.2018.12.080.

Tsubaki, J., & Jimbo, G. (1984). Theoretical analysis of the tensile strength of a powder bed. *Powder Technology, 37*(1), 219–227. https://doi.org/10.1016/0032-5910(84)80019-4.

Wu, Y., Du, X., Ge, H., & Lv, Z. (2011). Preparation of microporous starch by glucoamylase and ultrasound. *Starch - Stärke, 63*(4), 217–225. https://doi.org/10.1002/star.201000036.

Yan, H., Sun, E., Cui, L., Jia, X., & Jin, X. (2015). Improvement in oral bioavailability and dissolution of tanshinone IIA by preparation of solid dispersions with porous silica. *The Journal of Pharmacy and Pharmacology, 67*(9), 1207–1214. https://doi.org/10.1111/jphp.12423.

Zhang, B., Cui, D., Liu, M., Gong, H., Huang, Y., & Han, F. (2012). Corn porous starch: Preparation, characterization and adsorption property. *International Journal of Biological Macromolecules, 50*(1), 250–256. https://doi.org/10.1016/j.ijbiomac.2011.11.002.

Zhou, M., Shen, L., Lin, X., Hong, Y., & Feng, Y. (2017). Design and pharmaceutical applications of porous particles. *Royal Society of Chemistry (RSC), 7*(63), 39490–39501. https://doi.org/10.1039/c7ra06829h.

Chapter 8
Powders from Fruit Waste

Sahithi Murakonda and Madhuresh Dwivedi

8.1 Introduction

Fruits grown in geographical and sub-regions have been consumed all over the globe due to its numerous health benefits. The FAO and WHO recommends 400 g/day consumption of fruits to reduce the heart diseases, obesity, cancer and various health issues (FAO 2018). The great popularity of fruits has increased the annual production of the fruits in the global market subsequently advancing the processing technology. Apart from consuming as a whole fruit, the fruits due to its perishable nature are processed into many products like fruit juice, candy, jelly, powders etc. The industrial processing and inedible parts of fruits generates large amount of byproducts such as peel, seed etc., which is either disposed as fruit waste or used as fertilizer and cattle feed (Mirabella et al. 2014) in many industries and household. The fruit waste is of two kinds: (1) generated during sorting of bruised, spoiled fruits, dumping, transportation, and storage due to mechanical damage or spoilage of perishable nature (2) The inedible parts of fruits generated during processing (Shalini and Gupta 2010).

The food loss and waste by fruits and vegetables are higher than other commodities accounting 22% over the world production (FAO 2018). For example fruit waste produced: mango worldwide produces 75,000 Mt.; 10,384 Mt. of citrus fruit in Brazil; 26.5Mt in world, 39,000 t of Banana in US; 1 Mt. of apple pomance in India (Dorta et al. 2012; Bee Lin and Yek Cze 2018; Perazzini et al. 2013; Shalini and Gupta 2010). The disposal of fruit waste causes several environmental impacts as greenhouse gas (carbon dioxide, methane) emission, and diseases due to the rearing of flies, mice, microorganism around the fruit waste (Cheok et al. 2016).

Several fruits like banana, jackfruit, mango, pomegranate etc., comprises most portion as its inedible part, most of the nutritional portion containing in inedible part

S. Murakonda · M. Dwivedi (✉)
Department of Food Process Engineering, National Institute of Technology Rourkela, Rourkela, Odisha, India

E. Ermiş (ed.), *Food Powders Properties and Characterization*, Food Engineering Series, https://doi.org/10.1007/978-3-030-48908-3_8

is going as waste. The fruit waste powder arrests many health issues like cancer, cardiovascular, chronicle, diabetes, hypertension etc. (Dabas et al. 2013; Nascimento et al. 2016; Rawal and Masih 2014). Hence, several technologies and products have been developed to utilize complete nutritional benefits of the whole fruit.

The fruit waste containing several bioactive compounds by applying different processing techniques like drying, extraction etc., can be utilized as food product for human consumption apart from its other uses like biogas production, cattle feed and fertilizer for plants. The fruits waste to reduce the deterioration and to increase the shelf life and easy handling and reduce transportation cost is made into powders by removing its moisture content. The increase in the health consciousness of the human being has led to increase the demand of the nutritional availability in the food such as functional foods (Bhandari et al. 2013). The powders can be fortified with other products such as bakery, dairy, beverages etc., to increase the nutritional availability of other products or consumed as such.

This chapter furnishes the overview of the technologies developed for the extraction of bioactive compounds and drying some fruits waste for the development of powder and its fortified products.

8.2 Composition and Therapeutic Benefits of Fruit Waste

Banana peel containing 30–40% of its whole fruit weight is rich in 10–12% pectin, 6–12% lignin, 6–9% protein, 43.2–49.7% dietary fiber, 3% starch content (Castillo-Israel 2015; Khawas et al. 2015). The total phenolic and flavonoid content is more for the banana peel powder was more than the pulp powder (Fatemeh et al. 2012). Passion fruit is exotic fruit which contains 6–12% seed and 50–55% peel of fruit as byproducts. The peel is rich in 10–20 g pectin; dietary fiber;1.5 g protein; 56 g carbohydrates helps in preventing many diseases like cancer, cardiovascular issues, diabetes, diverticular, anti-hypertensiveness (Espírito Santo et al. 2012). Avocado byproducts are peel and 16% seeds which are rich in antioxidant activity than the edible portion of the fruit. The avocado seeds powder is good in reduction of diabetes, hypertension and boost hyper-cholesterolemia (Chel-Guerrero et al. 2016; Dabas et al. 2013). Orange peel constitute about 30–40% of the fruit. The wastewater from orange juice is good source of fiber (Manjarres-Pinzon et al. 2013; Mirabella et al. 2014). Dragon fruit peel has higher antioxidant property (Chia and Chong 2015). The prickly pear peel constitutes 40–50% of fruit and has protein (8.3%), pectin (3%), minerals (12.13%) (Lahsasni et al. 2004). The apple pomance is the byproduct (25%) of the apple produced during processing constitute seeds, peel and the rejected spoiled apples contains polyphenols, minerals (calcium 8.7%), carbohydrates(9.5–22%), protein (4%) and more dietary fiber than the oat and wheat bran which is resistant against cancer, chronicle issues (Mirabella et al. 2014; Rawal and Masih 2014). Grape seeds 30–40% and skin has antioxidant and resistance against many diseases like diabetes, radio protective, antitheperglycemic, anti-inflammatory effect (Salim & Salina 2017). Pomegranate seed constitute 22%

Table 8.1 Fruit waste composition of fruits

Fruit	Fruit waste composition	Reference
Banana	30–40% peel	Castillo-Israel et al. (2015)
Mango	11–18% skin; 14–22% seed	Ajila et al. (2010)
Orange	30–40% peel	Manjarres-Pinzon et al. (2013)
Apple pomance	25% of skin, seed, residue	Rawal and Masih (2014)
Papaya	10–20% skin;10–20% seed	Lee et al. (2011)
Pomegranate	50% peel; 22% seed	Goula and Adamopoulos (2012)
Jack fruit	52–62% skin;8–10% seed	Cheok et al. (2016)
Grape	30–40% seed	Salim & Salina (2017)
Pineapple	29–40% skin	Banerjee et al. (2018), Choonut et al. (2014)
Avocado	16% seed	Dabas et al. (2013)
Durian	55–65% skin; 5–15% seed	Siriphanich (2011)
Passion fruit	6–12% seed; 50–55% peel	Espírito Santo et al. (2012)
Rambutan	37–62% skin; 4–9% seed	Issara etal. (2014)
Dragon fruit	22–44% skin; 2–4%	Cheok et al. (2016)
Mangosteen	60–65% skin; 6–11% seed	
Prickly pear	40–50% peel	Lahsasni et al. (2004)

of the fruit in which seed oil content (12–20%) is rich it has antioxidant, immune function, estrogen content inhibition of eiosanoid enzyme property (Goula and Adamopoulos 2012). The extract of the blueberry waste powder (15–55%) is rich in phenolic acid, anthocyanin and flavonoid (Waterhouse et al. 2017). The fruits by-products (papaya, mango, passion fruit, Surinam cherry, guava, soursop, cashew apple, acerola, pineapple) are rich source of bioactive compounds compared to the edible pulp part of fruit (Silva et al. 2014), remaining some fruit waste are mentioned in the (Table 8.1).

8.3 Drying Technology

The drying is main post-harvest technology to reduce the moisture content of fruit waste for the development of powders. The fruit waste containing 70–90% of the moisture content is to be reduced below 10% for the storage of the fruit waste powders (Mirabella et al. 2014). The powder is developed by drying it below 10% moisture content and grinding the dried peel by using different equipment like ball mill, grinder etc. and then sieving to particular particle size. The drying inactivates the enzymes responsible for deterioration and increases the shelf life and organoleptic properties of the product (Kudra and Mujumdar 2009).

8.3.1 Hot-Air Oven Drying

The drying is done by exposing the fruit waste to hot air at constant temperature and humidity (Khawas et al. 2015). Oven helps in controlling the drying conditions for the drying like airflow rate (Bhandari et al. 2013). The drying time ranges from (mostly 12–48 h) (Alkarkhi et al. 2011). The effective moisture diffusivity of banana peel powder has been found at 70 °C (Khawas et al. 2015).

The ultrasound assisted convective hot-air dryer has maintained the total phenolic content, antioxidant activity of the passion fruit dried peel (Nascimento et al. 2016). The 30% oil is present in passion fruit seeds which has antioxidant capacity. Pereira et al. (2017) has dried the passion fruit seeds and extracted oil from the seed powder having more yield for soxhlet method (23.68%) followed by ultrasound extraction (20.96%).

8.3.2 Solar Drying

The solar dryer is developed alternative to open sun drying as it protective method of drying compared to sun-drying. It is economically advantageous, low-energy process, non-contaminating source of energy and gives good quality of product and can be used on-farm. Many low-cost solar dryers have been developed by the researchers (Amer et al. 2010; Bennamoun and Belhamri 2003). The solar dryer uses the sun-radiations through solar panels for the drying purpose.

It has been used for production of many powders from fruit waste like apple pomance, mango peel powder, pricky peel powder (Lahsasni et al. 2004). The solar dried fruit waste powder like pomegranate peel is incorporated in food products for preparation products like yogurt (Sah et al. 2016).

8.3.3 Spray Drying

The feed solution entering is atomized in the chamber and atomized liquid droplets are evaporated by hot air in contact leading to production of powder. The stickiness of powder is the disadvantage of the spray dried powder (Patel et al. 2009; Vehring et al. 2007).

The blueberry waste powder, red pitaya powder, bordo grape powder etc., are developed by the spray-drier (Souza et al. 2015; Shofinita and Langrish 2014; Waterhouse et al. 2017). Bakar et al. (2012) and Edrisi Sormoli and Langrish (2016) have optimized condition for development of Red pitaya peel powder at inlet and outlet air temperature of 165 °C and 80 °C, respectively and for orange peel powder at 150 °C inlet temperature and 80 °C outlet temperature.

8.3.4 Freeze Drying

The freeze drying is done at low temperatures and pressure under freezing point that removes the water content by sublimation where phase change from solid to vapor take place. As it is low temperature process, most of the physiochemical properties of the product are retained compared to other process but the process is expensive (Caliskan and Dirim 2017; Reyes et al. 2011).

The manage peel and seed powder, dragon fruit peel powder, pomegranate peel powder has retained more properties than oven dried (Dorta et al. 2012; Liaotrakoon et al. 2011; Mphahlele et al. 2016). Morais et al. (2015) has developed powder of seven fruits waste (pineapple, banana, avocado, passion fruit, watermelon, melon, papaya) and compared the flavonoid and phenolic content of the powder which resulted that, total phenolic was more avocado peel powder and melon peel powder has more flavonoid. Can-Cauich et al. (2017) have developed freeze-dried fruit peel powder from 11 fruits (purple sugar apple, green sugar apple, green star apple, purple star apple, white sapote, black sapote, custard apple, ilama, sapodilla, mamoncillo, dragon fruit) has stated that green sugar apple, purple sugar apple peel powder has higher total phenolic, flavonoid, antioxidant activity and all the 11 peel powders are good source of bioactive compounds.

8.3.5 Microwave and Radio-Frequency Drying

The heat generated in the microwave and radio frequency is due to volumetric heating i.e. the alternative electromagnetic field cause ionic movement and dipole rotation within the product. This has good drying kinetics, quality product and time required drying is less compared to other drying techniques. The researchers has developed combined microwave, radiofrequency with other techniques like hot-air, vacuum, irradiation to reduce the economic cost and improve the quality (Guo and Zhu 2014; Mermelstein 2001; Salim & Salina 2017).

Varith et al. (2007) have developed Logan peel powder using microwave-hot air drying which reduced the drying time and energy consumption. The banana peel powder produced by microwave irradiation showed good results than freeze drying, oven drying and vacuum drying (Vu et al. 2016). Puraikalan (2018) has developed banana peel powder using radiofrequency dryer and developed extruded product and pasta by incorporation of peel powder which enhanced the nutritional quality of the product.

The commonly used drying techniques have been forecasted above for the development of powder. The drying techniques like vacuum, rotary, drum drying, fluidized bed drying has also been used for development of the fruit waste powders. Galaz et al. (2017) have developed pomegranate peel powder by using drum dryer in which phenolic content was not affected by drying. Henríquez et al. (2010) developed apple peel powder using drum-drying which was resulted as good dietary fiber and phenolic compounds powder. The citrus sinensis peel powder with more vitamin C and limonene content was obtained using fluidized bed dryer than the open sun dried powder (Tasirin et al. 2014). Nevertheless, there need of more techniques which is cost-effective, can retain quality of the product. (Silva et al. 2016) have produced acerola residue powder using rotary dryer in which powder produced by rotatory dryer has good quality and source of bioactive compounds.

8.4 Products Developed from Fruit Waste Powders

The dried powders are incorporated in the food products like extruded products, dairy, bakery, confectionary etc. Due to the rich physicochemical properties they are recognized as functional foods or nutritional supplement and can also be used as a substitute. These products enhance the physicochemical properties like phytochemical, antioxidant activity of the volarized product (Bhandari et al. 2013).

The apple pomance, raspberry pomance powders can be used in bakery products which helps in enhancing the fiber content and organoleptic properties of product (Mirabella et al. 2014). Apple pomance powder was used for preparation of bakery products like cookies, toffees, sauce which was as substitute of soya meal and source of dietary fiber and can also be used for flavoring in juice industry (Shalini and Gupta 2010). The passion fruit peel, pineapple peel, pomegranate powder can be incorporated in the dairy products like skim yoghurt, probiotic yoghurt which is helpful in reducing the fermentation time and improves the texture (Espírito Santo et al. 2012; El-Said et al. 2014; Sah et al. 2016). The avocado peel powder is incorporated in the tea and results showed good antioxidant and sensory property (Rotta et al. 2015). The extruded product macaroni, biscuits with increased nutritional quality and sensory properties was prepared by incorporating the mango peel powder (Ajila et al. 2010; Mirabella et al. 2014).

8.4.1 Products from Bioactive Compounds of Fruit Waste Powders

The dried and milled fruit waste powders are used for extraction of the bioactive compounds by different extraction techniques in some studies. There are many extraction techniques like solvent extraction, supercritical fluid extraction, pulsed-electric field assisted, microwave assisted, ultrasound extraction, enzyme extraction, subcritical fluid extraction etc. which gives good yield of the bioactive compound from the fruit waste powders (Cheok et al. 2016; Kumar et al. 2017). The extracted bioactive compounds of fruit waste powders are encapsulated or added for development of product.

In some studies, from the dried fruit waste powders like banana peel powder, avocado seed powder, apple pomance powder etc., and bioactive compounds like pectin are extracted and used as an additive, gelling agent in some products (Chel-Guerrero et al. 2016; Mirabella et al. 2014). Castillo-Israel et al. (2015) has extracted pectin using hydrochloric acid and citric acid from the banana peel powder and incorporated in the strawberry jam as gelling agent. The pectin is extracted from the passion fruit dried peel (8–9% moisture content) using extraction techniques like electric field, normal extraction by boiling water etc. (Bezerra et al. 2015; Oliveira et al. 2015; Kulkarni and Vijayanand 2010). The developed Avocado seed powder can be used as gelling agent in food products and packaging material by using extracted starch from the powder (Chel-Guerrero et al. 2016; Dabas et al. 2013). The pectin is extracted from mango peel using hydrocolloid extraction after hot air drying of peel and also can be extracted by microwave and sonication assisted extraction in some studies (Banerjee et al. 2018). The bioactive compounds and products developed from selective fruit waste powders are furnished in (Table 8.2).

8.5 Other Advantages of Fruit Waste Powders

There are a few other advantages other than the product development in which some advantages are discussed in this chapter.

The powders are used as adsorbent in water treatment for removing dyes; metals (copper etc.) which causes water pollution. The wood apple shell, grape fruit peels, orange peel, banana peel powder and watermelon rind powders etc. has been used for the waste water treatment (Chen et al. 2018; Liu et al. 2012; Munagapati et al. 2018; Temesgen et al. 2018).

The fruit waste powders can be used for energy generation. The apple fruit waste powder is used as fuel of energy generation like steam, biogas which helps in reduction of the cost of energy generation (fossil fuels) and disposal and also used in bio-transformation (Sunny-Roberts et al. 2004; Shalini and Gupta 2010). Nagle et al. (2011) has used peels, seed from litchi, mango, and logon as fuel for drying.

The fruit waste powders are used in packaging as bio-plastic films such as PHA (polyhydroxyalkanoates) in the green application instead of the degrading plastics which are toxic to the environment. The bio-plastic is better oxygen, moisture barrier properties. The fruit waste like citrus fruit waste, grape pomance, apricot waste etc., is used for preparation of bio-plastic films (Tsang et al. 2019; Yaradoddi et al. 2017). Moro et al. (2017) has developed bio-plastic with passion fruit peel powder containing 60% fiber.

The compounds like anthocyanin from fruit waste powder are used as natural food colorant and bio-pigment (Gunjal 2019). The pomegranate peel powder, red pithaya peel powder etc., are the natural source of colorant (Ajmal et al. 2014).

There are many other advantages from fruit waste powder like as fertilizer, cattle feed, emulsifiers, essential oil, preservative, ethanol production etc. (Choonut et al. 2014; Gunjal 2019; Khedkar et al. 2017).

Table 8.2 Bioactive compounds and products developed from fruit waste powder

Fruit	Fruit Waste Powders	Bioactive compounds	Product developed	Reference
Banana	Banana peel powder	Anthocyanin, pectin, cyaniding, gallocatechin etc.	Gelly, bakery products, strawberry jam etc.	Castillo-Israel et al. (2015), González-Montelongo et al. (2010), Lee et al. (2010)
Apple	Apple pomance powder	Chlorogenic acid, phloretin, catechins, anthocynins, glycosides etc.	Cookies, toffee, sauce etc.	Shalini and Gupta (2010), Wolfe and Liu (2003)
Pomegranate	Pomegranate peel powder	Gallic acid, cyaniding etc.,	Dairy product like yogurt	El-Said et al. (2014)
Avocado	Avocado peel powder	Chlorogenic acid, catechin, gallic acid etc.	Tea	Rotta et al. (2015)
Mango	Mango peel powder	Carotenoid, anthocyanin, polyphenols etc.	Biscuits, macaroni	Ajila et al. (2010)
Orange	Orange peel powder	Limonene, linalool, polyphenols	Wheat flour bun, biscuits	Kumar et al. (2017), Raj and Masih (2014), Youssef and Mousa (2012)
Grape	Grape skin, seed powder	Caffeic acid, gallic acid, cinnamic acid, ferulic acid, vanillic acid, anthocyanins, proanthocyanidins etc.	Fruit candies, chicken patties and nuggets	Cagdas and Kumcuoglu (2015), Cappa et al. (2015), Maier et al. (2009), Ruiz-Capillas et al. (2017)
Apricot	Apricot seed powder	Carotenoids, β-carotene etc	Bakery products like cookies	Kumar et al. (2017), Özboy-Özbaş et al. (2010)
Passion fruit	Passion fruit peel powder	Pectin, β-carotene etc.	Probiotic product like yogurt	Espírito Santo et al. (2012), Oliveira et al. (2016)
Guava	Guava peel powder	Gallic acid, kaempferol, galangin, homogentisic acid etc.	Bakery products-cookies	Bertagnolli et al. (2014), Deng et al. (2012)

8.6 Conclusion

In world, greater than one billion tons of fruit waste is produced as most waste is going in the inedible part and also due to perishable nature and mechanical damage of fruits. The inedible part of some fruit contains more bioactive compounds and health benefits than edible part which is disposed in many industries causing environmental issues. These fruit waste are converted into edible ingredient by processing into powders by reducing its moisture content to increase the shelflife and to inactivate enzymes responsible for deterioration. Many processing technologies like drying, extraction etc. have been developed for efficient utilization of the fruit waste. These powders have been incorporated into many food products like bakery, confectionary, and dairy products to increase functional properties of the developed product. The rich source of bioactive compounds present in fruit waste powders made to develop many extraction techniques for extraction of the bioactive compounds and volarizing for new product.

Even though, there is development of many techniques, many industries are still disposing the fruit waste or using as cattle feed. Every fruit industry should make an attempt and start for protecting the environment by utilizing the fruit waste for therapeutic human consumption instead of dumping it. Nevertheless, advanced techniques in drying has been developed, still these is issue like stickiness, nutrient loss and rehydration of moisture in some fruit powders. Hence, further research should be undertaken by researcher to overcome the problems. Many techniques for incorporating bioactive compounds other than encapsulation, co-crystallization etc. need to be developed, as some compounds inactivate in certain environment as they are not stable. Hence, there is more scope for research in the development of fruit waste powders. The fruit waste powder is beneficial in environmental, therapeutic and economic point of view.

References

Ajmal, M., Adeel, S., Azeem, M., Zuber, M., Akhtar, N., & Iqbal, N. (2014). Modulation of pomegranate peel colourant characteristics for textile dyeing using high energy radiations. *Industrial Crops and Products, 58*, 188–193. https://doi.org/10.1016/j.indcrop.2014.04.026.

Ajila, C. M., Aalami, M., Leelavathi, K., & Rao, U. J. S. P. (2010). Mango peel powder: A potential source of antioxidant and dietary fiber in macaroni preparations. *Innovative Food Science & Emerging Technologies, 11*(1), 219–224.

Alkarkhi, A. F. M., Ramli, S. b., Yong, Y. S., & Easa, A. M. (2011). Comparing physicochemical properties of banana pulp and peel flours prepared from green and ripe fruits. *Food Chemistry, 129*(2), 312–318. https://doi.org/10.1016/j.foodchem.2011.04.060.

Amer, B. M. A., Hossain, M. A., & Gottschalk, K. (2010). Design and performance evaluation of a new hybrid solar dryer for banana. *Energy Conversion and Management, 51*(4), 813–820. https://doi.org/10.1016/j.enconman.2009.11.016.

Bakar, J., Ee, S. C., Muhammad, K., Hashim, D. M., & Adzahan, N. (2012). Spray-drying optimization for red pitaya Peel (Hylocereus polyrhizus). *Food and Bioprocess Technology, 6*(5), 1332–1342. https://doi.org/10.1007/s11947-012-0842-5.

Banerjee, J., Singh, R., Vijayaraghavan, R., MacFarlane, D., Patti, A. F., & Arora, A. (2018). A hydrocolloid based biorefinery approach to the valorisation of mango peel waste. *Food Hydrocolloids, 77*, 142–151. https://doi.org/10.1016/j.foodhyd.2017.09.029.

Bee Lin, C., & Yek Cze, C. (2018). Drying kinetics and optimisation of pectin extraction from Banana peels via response surface methodology. *MATEC Web of Conferences, 152*, 1002. https://doi.org/10.1051/matecconf/201815201002.

Bennamoun, L., & Belhamri, A. (2003). Design and simulation of a solar dryer for agriculture products. *Journal of Food Engineering, 59*(2–3), 259–266. https://doi.org/10.1016/s0260-8774(02)00466-1.

Bertagnolli, S. M. M., Silveira, M. L. R., de Fogaça, A. O., Umann, L., & Penna, N. G. (2014). Bioactive compounds and acceptance of cookies made with Guava peel flour. *Food Science and Technology (Campinas), 34*(2), 303–308.

Bezerra, C. V., Meller da Silva, L. H., Corrêa, D. F., & Rodrigues, A. M. C. (2015). A modeling study for moisture diffusivities and moisture transfer coefficients in drying of passion fruit peel. *International Journal of Heat and Mass Transfer, 85*, 750–755. https://doi.org/10.1016/j.ijheatmasstransfer.2015.02.027.

Bhandari, B., Bansal, N., Zhang, M., & Schuck, P. (2013). *Handbook of food powders*. Oxford: Woodhead Publishing. https://doi.org/10.1533/9780857098672.

Cappa, C., Lavelli, V., & Mariotti, M. (2015). Fruit candies enriched with grape skin powders: Physicochemical properties. *LWT - Food Science and Technology, 62*(1), 569–575.

Cagdas, E., & Kumcuoglu, S. (2015). Effect of grape seed powder on oxidative stability of precooked chicken nuggets during frozen storage. *Journal of Food Science and Technology, 52*(5), 2918–2925.

Caliskan, G., & Dirim, S. N. (2017). Drying characteristics of pumpkin (Cucurbita moschata) slices in convective and freeze dryer. *Heat and Mass Transfer, 53*(6), 2129–2141. https://doi.org/10.1007/s00231-017-1967-x.

Castillo-Israel, S. F., Baguio, M. D. B., Diasanta, R. C. M., Lizardo, E. I., & Dizon, M. M. (2015). Extraction and characterization of pectin from Saba banana. *International Food Research Journal, 22*(1), 202–207.

Can-Cauich, C. A., Sauri-Duch, E., Betancur-Ancona, D., Chel-Guerrero, L., González-Aguilar, G. A., Cuevas-Glory, L. F., et al. (2017). Tropical fruit peel powders as functional ingredients: Evaluation of their bioactive compounds and antioxidant activity. *Journal of Functional Foods, 37*, 501–506. https://doi.org/10.1016/j.jff.2017.08.028.

Chel-Guerrero, L., Barbosa-Martín, E., Martínez-Antonio, A., González-Mondragón, E., & Betancur-Ancona, D. (2016). Some physicochemical and rheological properties of starch isolated from avocado seeds. *International Journal of Biological Macromolecules, 86*, 302–308. https://doi.org/10.1016/j.ijbiomac.2016.01.052.

Chen, Y., Wang, H., Zhao, W., & Huang, S. (2018). Four different kinds of peels as adsorbents for the removal of Cd (II) from aqueous solution: Kinetics, isotherm and mechanism. *Journal of the Taiwan Institute of Chemical Engineers, 88*, 146–151. https://doi.org/10.1016/j.jtice.2018.03.046.

Cheok, C. Y., Mohd Adzahan, N., Abdul Rahman, R., Zainal Abedin, N. H., Hussain, N., Sulaiman, R., & Chong, G. H. (2016). Current trends of tropical fruit waste utilization. *Critical Reviews in Food Science and Nutrition, 58*, 1–27. https://doi.org/10.1080/10408398.2016.1176009.

Chia, S. L., & Chong, G. H. (2015). Effect of drum drying on physico-chemical characteristics of dragon fruit Peel (Hylocereus polyrhizus). *International Journal of Food Engineering, 11*(2), 285–293. https://doi.org/10.1515/ijfe-2014-0198.

Choonut, A., Saejong, M., & Sangkharak, K. (2014). The production of ethanol and hydrogen from pineapple peel by saccharomyces cerevisiae and Enterobacter Aerogenes. *Energy Procedia, 52*, 242–249. https://doi.org/10.1016/j.egypro.2014.07.075.

da Silva, L. M. R., de Figueiredo, E. A. T., Ricardo, N. M. P. S., Vieira, I. G. P., de Figueiredo, R. W., Brasil, I. M., & Gomes, C. L. (2014). Quantification of bioactive compounds in pulps and by-products of tropical fruits from Brazil. *Food Chemistry, 143*, 398–404. https://doi.org/10.1016/j.foodchem.2013.08.001.

Dabas, D., Shegog, R., Ziegler, G., & Lambert, J. (2013). Avocado (Persea americana) seed as a source of bioactive phytochemicals. *Current Pharmaceutical Design, 19*(34), 6133–6140. https://doi.org/10.2174/13816128113193400007.

Deng, G.-F., Shen, C., Xu, X.-R., Kuang, R.-D., Guo, Y.-J., Zeng, L.-S., et al. (2012). Potential of fruit wastes as natural resources of bioactive compounds. *International Journal of Molecular Sciences, 13*(7), 8308–8323.

Dorta, E., Lobo, M. G., & González, M. (2012). Using drying treatments to stabilise mango peel and seed: Effect on antioxidant activity. *LWT - Food Science and Technology, 45*(2), 261–268. https://doi.org/10.1016/j.lwt.2011.08.016.

Espírito Santo, A. P. do, Perego, P., Converti, A., & Oliveira, M. N. (2012). Influence of milk type and addition of passion fruit peel powder on fermentation kinetics, texture profile and bacterial viability in probiotic yoghurts. *LWT, 47*(2), 393–399. https://doi.org/10.1016/j.lwt.2012.01.038.

Edrisi Sormoli, M., & Langrish, T. A. G. (2016). Spray drying bioactive orange-peel extracts produced by Soxhlet extraction: Use of WPI, antioxidant activity and moisture sorption isotherms. *LWT - Food Science and Technology, 72*, 1–8. https://doi.org/10.1016/j.lwt.2016.04.033.

El-Said, M. M., Haggag, H. F., Fakhr El-Din, H. M., Gad, A. S., & Farahat, A. M. (2014). Antioxidant activities and physical properties of stirred yoghurt fortified with pomegranate peel extracts. *Annals of Agricultural Sciences, 59*(2), 207–212. https://doi.org/10.1016/j.aoas.2014.11.007.

FAO (2018). SDG 12.3.1 Global food loss index. http://www.fao.org/3/CA2640EN/ca2640en.pdf. Access date: 10.02.2019.

Fatemeh, S. R., Saifullah, R., Abbas, F. M. A., & Azhar, M. E. (2012). Total phenolics, flavonoids and antioxidant activity of banana pulp and peel flours: Influence of variety and stage of ripeness. *International Food Research Journal, 19*(3), 1041–1046.

Galaz, P., Valdenegro, M., Ramírez, C., Nuñez, H., Almonacid, S., & Simpson, R. (2017). Effect of drum drying temperature on drying kinetic and polyphenol contents in pomegranate peel. *Journal of Food Engineering, 208*, 19–27. https://doi.org/10.1016/j.jfoodeng.2017.04.002.

González-Montelongo, R., Gloria Lobo, M., & González, M. (2010). Antioxidant activity in banana peel extracts: Testing extraction conditions and related bioactive compounds. *Food Chemistry, 119*(3), 1030–1039.

Goula, A. M., & Adamopoulos, K. G. (2012). A method for pomegranate seed application in food industries: Seed oil encapsulation. *Food and Bioproducts Processing, 90*(4), 639–652. https://doi.org/10.1016/j.fbp.2012.06.001.

Gunjal, B. B., (2019). Value-added products from food waste. *Advances in Environmental Engineering and Green Technologies*, pp.20–30. https://doi.org/10.4018/978-1-5225-7706-5.ch002

Guo, W., & Zhu, X. (2014). Dielectric properties of red pepper powder related to radiofrequency and microwave drying. *Food and Bioprocess Technology, 7*(12), 3591–3601. https://doi.org/10.1007/s11947-014-1375-x.

Henríquez, C., Speisky, H., Chiffelle, I., Valenzuela, T., Araya, M., Simpson, R., & Almonacid, S. (2010). Development of an ingredient containing apple peel, as a source of polyphenols and dietary Fiber. *Journal of Food Science, 75*(6), H172–H181. https://doi.org/10.1111/j.1750-3841.2010.01700.x.

Issara, U., Zzaman, W., & Yang, T. A. (2014). Rambutan seed fat as a potential source of cocoa butter substitute in confectionary product. *International Food Research Journal, 21*(1), 25–31.

Khawas, P., Dash, K. K., Das, A. J., & Deka, S. C. (2015). Drying characteristics and assessment of physicochemical and microstructural properties of dried culinary banana slices. *International Journal of Food Engineering, 11*(5), 667–678. https://doi.org/10.1515/ijfe-2015-0094.

Khedkar, M. A., Nimbalkar, P. R., Gaikwad, S. G., Chavan, P. V., & Bankar, S. B. (2017). Sustainable biobutanol production from pineapple waste by using clostridium acetobutylicum B 527: Drying kinetics study. *Bioresource Technology, 225*, 359–366. https://doi.org/10.1016/j.biortech.2016.11.058.

Kudra, T., & Mujumdar, A. S. (2009). *Advanced drying technologies*. Boca Raton: CRC Press. https://doi.org/10.1201/9781420073898.

Kulkarni, S. G., & Vijayanand, P. (2010). Effect of extraction conditions on the quality characteristics of pectin from passion fruit peel (Passiflora edulis f. flavicarpa L.). *LWT - Food Science and Technology, 43*(7), 1026–1031. https://doi.org/10.1016/j.lwt.2009.11.006.

Kumar, K., Yadav, A. N., Kumar, V., Vyas, P., & Dhaliwal, H. S. (2017). Food waste: A potential bioresource for extraction of nutraceuticals and bioactive compounds. *Bioresources and Bioprocessing, 4*(1).

Lahsasni, S., Kouhila, M., Mahrouz, M., Idlimam, A., & Jamali, A. (2004). Thin layer convective solar drying and mathematical modeling of prickly pear peel (Opuntia ficus indica). *Energy, 29*(2), 211–224. https://doi.org/10.1016/j.energy.2003.08.009.

Lee, E.-H., Yeom, H.-J., Ha, M.-S., & Bae, D.-H. (2010). Development of banana peel jelly and its antioxidant and textural properties. *Food Science and Biotechnology, 19*(2), 449–455.

Lee, W.-J., Lee, M.-H., & Su, N.-W. (2011). Characteristics of papaya seed oils obtained by extrusion-expelling processes. *Journal of the Science of Food and Agriculture, 91*(13), 2348–2354.

Liaotrakoon, W., De Clercq, N., Van Hoed, V., Van de Walle, D., Lewille, B., & Dewettinck, K. (2011). Impact of thermal treatment on physicochemical, antioxidative and rheological properties of white-flesh and red-flesh dragon fruit (Hylocereus spp.) purees. *Food and Bioprocess Technology, 6*(2), 416–430. https://doi.org/10.1007/s11947-011-0722-4.

Liu, C., Ngo, H. H., Guo, W., & Tung, K.-L. (2012). Optimal conditions for preparation of banana peels, sugarcane bagasse and watermelon rind in removing copper from water. *Bioresource Technology, 119*, 349–354. https://doi.org/10.1016/j.biortech.2012.06.004.

Maier, T., Schieber, A., Kammerer, D. R., & Carle, R. (2009). Residues of grape (Vitis vinifera L.) seed oil production as a valuable source of phenolic antioxidants. *Food Chemistry, 112*(3), 551–559.

Manjarres-Pinzon, K., Cortes-Rodriguez, M., & Rodríguez-Sandoval, E. (2013). Effect of drying conditions on the physical properties of impregnated orange peel. *Brazilian Journal of Chemical Engineering, 30*(3), 667–676. https://doi.org/10.1590/s0104-66322013000300023.

Mermelstein, N. H. (2001). Spray drying. *Food Technology, 55*(4), 92–95.

Mirabella, N., Castellani, V., & Sala, S. (2014). Current options for the valorization of food manufacturing waste: A review. *Journal of Cleaner Production, 65*, 28–41. https://doi.org/10.1016/j.jclepro.2013.10.051.

Morais, D. R., Rotta, E. M., Sargi, S. C., Schmidt, E. M., Bonafe, E. G., Eberlin, M. N., et al. (2015). Antioxidant activity, phenolics and UPLC–ESI(–)–MS of extracts from different tropical fruits parts and processed peels. *Food Research International, 77*, 392–399. https://doi.org/10.1016/j.foodres.2015.08.036.

Moro, T. M. A., Ascheri, J. L. R., Ortiz, J. A. R., Carvalho, C. W. P., & Meléndez-Arévalo, A. (2017). Bioplastics of native starches reinforced with passion fruit Peel. *Food and Bioprocess Technology, 10*(10), 1798–1808. https://doi.org/10.1007/s11947-017-1944-x.

Mphahlele, R. R., Fawole, O. A., Makunga, N. P., & Opara, U. L. (2016). Effect of drying on the bioactive compounds, antioxidant, antibacterial and antityrosinase activities of pomegranate peel. *BMC Complementary and Alternative Medicine, 16*, 143. Available at: https://pubmed.ncbi.nlm.nih.gov/27229852.

Munagapati, V. S., Yarramuthi, V., Kim, Y., Lee, K. M., & Kim, D.-S. (2018). Removal of anionic dyes (reactive black 5 and Congo red) from aqueous solutions using Banana Peel powder as an adsorbent. *Ecotoxicology and Environmental Safety, 148*, 601–607. https://doi.org/10.1016/j.ecoenv.2017.10.075.

Nagle, M., Habasimbi, K., Mahayothee, B., Haewsungcharern, M., Janjai, S., & Müller, J. (2011). Fruit processing residues as an alternative fuel for drying in Northern Thailand. *Fuel, 90*(2), 818–823. https://doi.org/10.1016/j.fuel.2010.10.003.

Nascimento, E. M. G. C. do, Mulet, A., Ascheri, J. L. R., de Carvalho, C. W. P., & Cárcel, J. A. (2016). Effects of high-intensity ultrasound on drying kinetics and antioxidant properties of passion fruit peel. *Journal of Food Engineering, 170*, 108–118. https://doi.org/10.1016/j.jfoodeng.2015.09.015.

Oliveira, C. F., Giordani, D., Gurak, P. D., Cladera-Olivera, F., & Marczak, L. D. F. (2015). Extraction of pectin from passion fruit peel using moderate electric field and conventional heating extraction methods. *Innovative Food Science & Emerging Technologies, 29*, 201–208. https://doi.org/10.1016/j.ifset.2015.02.005.

Oliveira, C. F., Gurak, P. D., Cladera-Olivera, F., & Marczak, L. D. F. (2016). Evaluation of physicochemical, technological and morphological characteristics of powdered yellow passion fruit peel. *International Food Research Journal, 23*(4), 1653–1662.

Özboy-Özbaş, Ö., Seker, I. T., & Gökbulut, I. (2010). Effects of resistant starch, apricot kernel flour, and fiber-rich fruit powders on low-fat cookie quality. *Food Science and Biotechnology, 19*(4), 979–986.

Patel, R. P., Patel, M. P., & Suthar, A. M. (2009). Spray drying technology: An overview. *Indian Journal of Science and Technology, 2*(10), 44–47.

Perazzini, H., Freire, F. B., & Freire, J. T. (2013). Drying kinetics prediction of solid waste using semi-empirical and artificial neural network models. *Chemical Engineering & Technology, 36*(7), 1193–1201. https://doi.org/10.1002/ceat.201200593.

Pereira, M. G., Hamerski, F., Andrade, E. F., de Scheer, A. P., & Corazza, M. L. (2017). Assessment of subcritical propane, ultrasound-assisted and Soxhlet extraction of oil from sweet passion fruit (Passiflora alata Curtis) seeds. *The Journal of Supercritical Fluids, 128*, 338–348. https://doi.org/10.1016/j.supflu.2017.03.021.

Puraikalan, Y. (2018). Characterization of proximate, phytochemical and antioxidant analysis of Banana (Musa Sapientum) peels/skins and objective evaluation of ready to eat /cook product made with banana peels. *Current Research in Nutrition and Food Science Journal, 6*(2), 382–391. https://doi.org/10.12944/crnfsj.6.2.13.

Raj, A., & Masih, D. (2014). Physico chemical and rheological properties of wheat flour bun supplemented with orange peel powder. *International Journal of Science and Research, 3*(8), 391–394.

Rawal, R., & Masih, D. (2014). Study of the effect on the quality attributes of apple pomace powder prepared by two different dryers. *IOSR Journal of Agriculture and Veterinary Science, 7*(8), 54–61. https://doi.org/10.9790/2380-07825461.

Reyes, A., Mahn, A., & Huenulaf, P. (2011). Drying of apple slices in atmospheric and vacuum freeze dryer. *Drying Technology, 29*(9), 1076–1089. https://doi.org/10.1080/07373937.2011.568657.

Rotta, E. M., de Morais, D. R., Biondo, P. B. F., dos Santos, V. J., Matsushita, M., & Visentainer, J. V. (2015). Use of avocado peel (Persea americana) in tea formulation: A functional product containing phenolic compounds with antioxidant activity. *Acta Scientiarum Technology, 38*(1), 23. Available at: http://periodicos.uem.br/ojs/index.php/ActaSciTechnol/article/view/27397.

Ruiz-Capillas, C., Nardoia, M., Herrero, A. M., Jimnez-Colmenero, F., Chamorro, S., & Brenes, A. (2017). Effect of added grape seed and skin on chicken thigh patties during chilled storage. *International Journal of Food and Nutritional Science, 4*(1), 67–73.

Sah, B. N. P., Vasiljevic, T., McKechnie, S., & Donkor, O. N. (2016). Effect of pineapple waste powder on probiotic growth, antioxidant and antimutagenic activities of yogurt. *Journal of food science and technology, 53*(3), 1698–1708. Available at: https://pubmed.ncbi.nlm.nih.gov/27570295.

Salim, M. D., & Salina, N. (2017). Potential utilization of fruit and vegetable wastes for food through drying or extraction techniques. *Novel Techniques in Nutrition & Food Science, 1*(2). https://doi.org/10.31031/ntnf.2017.01.000506.

Shalini, R., & Gupta, D. K. (2010). Utilization of pomace from apple processing industries: A review. *Journal of food science and technology, 47*(4), 365–371. Available at: https://pubmed.ncbi.nlm.nih.gov/23572655.

Shofinita, D., & Langrish, T. A. G. (2014). Spray drying of orange peel extracts: Yield, total phenolic content, and economic evaluation. *Journal of Food Engineering, 139*, 31–42. https://doi.org/10.1016/j.jfoodeng.2014.03.028.

Silva, P. B., Duarte, C. R., & Barrozo, M. A. S. (2016). Dehydration of acerola (Malpighia emarginata D.C.) residue in a new designed rotary dryer: Effect of process variables on main bioactive compounds. *Food and Bioproducts Processing, 98*, 62–70. https://doi.org/10.1016/j.fbp.2015.12.008.

Souza, V. B., Thomazini, M., & Balieiro, J. C. de C., & Fávaro-Trindade, C. S. (2015). Effect of spray drying on the physicochemical properties and color stability of the powdered pigment obtained from vinification byproducts of the Bordo grape (Vitis labrusca). *Food and Bioproducts Processing, 93*, 39–50. https://doi.org/10.1016/j.fbp.2013.11.001.

Siriphanich, J. (2011). Durian (*Durio zibethinus Merr.*). In E. Yahia (Ed.), *Postharvest biology and Technology of Tropical and Subtropical Fruits* (Woodhead publishing series in food science). Woodhead Publishing.

Sunny-Roberts, E. O., Otunola, E. T., & Iwakun, B. T. (2004). An evaluation of some quality parameters of a laboratory-prepared fermented groundnut milk. *European Food Research and Technology, 218*(5), 452–455.

Tasirin, S. M., Puspasari, I., Sahalan, A. Z., Mokhtar, M., Ghani, M. K. A., & Yaakob, Z. (2014). Drying ofCitrus sinensisPeels in an inert fluidized bed: Kinetics, microbiological activity, vitamin C, and limonene determination. *Drying Technology, 32*(5), 497–508. https://doi.org/10.1080/07373937.2013.838782.

Temesgen, F., Gabbiye, N., & Sahu, O. (2018). Biosorption of reactive red dye (RRD) on activated surface of banana and orange peels: Economical alternative for textile effluent. *Surfaces and Interfaces, 12*, 151–159. https://doi.org/10.1016/j.surfin.2018.04.007.

Tsang, Y. F., Kumar, V., Samadar, P., Yang, Y., Lee, J., Ok, Y. S., et al. (2019). Production of bioplastic through food waste valorization. *Environment International, 127*, 625–644. https://doi.org/10.1016/j.envint.2019.03.076.

Varith, J., Dijkanarukkul, P., Achariyaviriya, A., & Achariyaviriya, S. (2007). Combined microwave-hot air drying of peeled longan. *Journal of Food Engineering, 81*(2), 459–468. https://doi.org/10.1016/j.jfoodeng.2006.11.023.

Vehring, R., Foss, W. R., & Lechuga-Ballesteros, D. (2007). Particle formation in spray drying. *Journal of Aerosol Science, 38*(7), 728–746. https://doi.org/10.1016/j.jaerosci.2007.04.005.

Vu, H. T., Scarlett, C. J., & Vuong, Q. V. (2016). Effects of drying conditions on physicochemical and antioxidant properties of banana (Musa cavendish) peels. *Drying Technology, 35*(9), 1141–1151. https://doi.org/10.1080/07373937.2016.1233884.

Waterhouse, G. I. N., Dijkanarukkul, P., Achariyaviriya, A., & Achariyaviriya, S. (2017). Spray-drying of antioxidant-rich blueberry waste extracts; interplay between waste pretreatments and spray-drying process. *Food and Bioprocess Technology, 10*(6), 1074–1092. https://doi.org/10.1007/s11947-017-1880-9.

Wolfe, K. L., & Liu, R. H. (2003). Apple peels as a value-added food ingredient. *Journal of Agricultural and Food Chemistry, 51*(6), 1676–1683.

Yaradoddi, J. S., Hugar, S., Banapurmath, N. R., Hunashyal, A. M., Sulochana, M. B., Shettar, A. S., & Ganachari, S. V. (2017). Alternative and renewable bio-based and biodegradable plastics. In *Handbook of ecomaterials* (pp. 1–20). Cham: Springer. https://doi.org/10.1007/978-3-319-48281-1_150-1.

Youssef, H., & Mousa, R. (2012). Nutritional assessment of wheat biscuits and fortified wheat biscuits with citrus peels powders. *Food and Public Health, 2*(1), 55–60.

Chapter 9
The Microbiological Safety of Food Powders

E. J. Rifna and Madhuresh Dwivedi

9.1 Introduction

The development of formulation concept within food industries and insist intended for multiplicity in food goods have boosted an efficient market advancement for food ingredients. Preponderance of products is formed in pulverized state, and consequently powder technology is an increasingly high imperative issue equally to food component manufacturers and food producers. The chief basis for powdered foods to signify the major fraction of the whole prepackaged food is their low water activity, thus, dropping the pace of quality deprivation. Low moisture foods are produced from high moisture foods which are frequently subjected to desiccation or dehydration processes to accomplish this, or they possibly will be physically low in moisture content. According to the definition of Food and Agricultural Organization (FAO), low low water activity (a_w) products are the food samples with water activity in range of less than 0.7 (Blessington et al. 2013). Globally, a broad range of food produce are mechanically transformed to fine granular state owing to its extended utility and stability varying from grain flours, cereal flours, dried egg powder, dried milk powder, dried spice and herbs, PIF, culinary powder, fruit or vegetable powder.

As stated above food powders being low in water activity and residual moisture content, they retard the development of pathogenic and spoilage microbes thus making product safe from food borne pathogens. In addition, there is a general misapprehension that microorganisms are not able to endure and develop in food products with low moisture content. Researches has revealed that food borne bacteria's, including *Salmonella*, *Cronobacter*, *Clostridium* and *Bacillus* grow and proliferate in dried powdered foods (Chitrakar et al. 2018) thus initiating to be the origin of microbial outbreaks in food powders. Epidemiological documentation inferred from these breakouts depicted that probable cross contamination could have played

E. J. Rifna · M. Dwivedi (✉)
Department of Food Process Engineering, National Institute of Technology Rourkela, Rourkela, Odisha, India

E. Ermiş (ed.), *Food Powders Properties and Characterization*, Food Engineering Series, https://doi.org/10.1007/978-3-030-48908-3_9

a significant role in food powder contamination, even the Salmonella breakout in India leading to recall of sambhar masala in year 2019 from 'House of Spices'. It has been projected that 25% of food-borne breakouts in food powders was through cross contamination (Reilly et al. 2009). Cross-contamination can consequences from insufficient hygienic implementation, lack of precise awareness of sanitation in food handling, contaminated equipment, or inadequate storage surroundings. To add, it has also been understood that the majority food powders are not in a straight line ingested by consumers and be generally mixed using liquids majorly water to create wet formulations, that are as such consumed. It has been premeditated through a numerous of studies that reconstitution of powder foods carrying microbial facilitated cells to commence damage repair developing end food dangerous for consumption. Presently, a small number of fatal breakouts of food-borne pathogens came into sight respect with powdered foods during recombination; *Cronobacter sakazakii* in infant formula, *Salmonella* in milk powder, *Bacillus aureus* in vegetable powder (Endersen et al. 2017; Juneja et al. 2017; Wang et al. 2020), leading to number of clinical cases among consumers. As more frequent exposition of food-borne outbreaks allied to powder foods are recorded worldwide, the food safety rank of the universally assumed safe "powder foods" is attaining more interest at the moment. This chapter presents the major outstanding results attained so far considering the microbiological safety in food powders. The microbial safety issues in different food powders together with assortment of food borne outbreaks, and effect of emerging non-thermal techniques for microbial decontamination of powdered foods are discussed here.

9.2 Implicated Food Borne Pathogens Linked with Powdered Foods

Historically, abundant foodborne diseases have been related with food powders. In the precedent decade, nevertheless, the occurrence of these outbreaks has amplified, probably owing to novel responsive and robust microbiological testing techniques, enhanced epidemiologic traceback tracking methods, and highly proficient sampling process (Gurtler et al. 2014). Though a varied amount of foodborne microbes have been isolated out to contaminate food powders, a significant count of *Salmonella* serovars have been the recurrently indicated bacteria, when associated to other members of bacilli family (*E. coli*, *Shigella*, *Lactobacillus*, *Listeria, Staphyllococcus* and *Listeria monocytogenes*). However, one anomaly is the bacilli genes *Cronobacter sakazakii* (formerly known as *Enterobacter sakazakii*), which has resulted in exceptional but frequent rigorous cases of illness in immunosuppressed adults and neonates. Further, though certain spore forming microorganisms are not able to survive in low moisture condition (eg: *Clostridium, Bacillus*) on rehydration, survivors may present a foodborne illness risk through production of toxins and toxico-infections. The subsequent text is intended to present added insights appropriate to the characteristics of pathogens that have been linked with, or recognized to have triggered breakouts of foodborne illness owing to consumption of contaminated food powders.

9.2.1 *Salmonella*

Salmonellae are a genus of Gram-negative, rod shaped and non-sporing bacteria belonging to family *Enterobacteriaceae*. The main two kinds of *Salmonella* are *Salmonella bongori* and *Salmonella enteric* (Abatcha et al. 2014). The inexplicably superior figure of *Salmonella* –associated outbreaks and recalls associated with food powders may be correlated to the remarkable capability of *Salmonella* species to endure dehydrated and/or low moisture environment, when compared to further members of the *Enterobacteriaceae* bacilli family or when compared to further non-sporing bacilli. Further, in food powders as the a_w is decreased by solute addition or by water removal, resistance to heat enhances proficiently. In foods powders such as diary powders and spices several minutes at temperature above 110 °C may be vital to diminish *Salmonella* counts to 1 log cfu/g (Podolak et al. 2010). For powders containing low levels of fat and/or protein there is an elevated possibility of outbreaks at doses of >105 cells however in food powders holding elevated levels of fat/ protein, such as cottage cheese powder, egg powder, milk powder, lentil powder, whey protein powder and soy powder, contamination can upshot from consuming as few as <10–100 cells. Globally, there are a projected 1.3 billion cases of gastroenteritis caused due to Salmonella infection, leading to about three million deaths every year worldwide. *Salmonella* infection also triggers mortality, morbidity and nausea across the globe with the immune response variegating relying on whether the infection is chronic, acute, or limited (Humphries and Linscott 2015). Breakouts of salmonella related foodborne illness are majorly caused by insufficient manage of temperatures at cooking, cross-contamination post to cooking, decreased cooling rates, and meager refrigeration.

Numerous parameters such as initial load of microorganisms, powder type, powder water activity and product temperature were reported to effect on viability of salmonella species in dried food powders. Hu et al. (2017) deduced that *Salmonella* Typhimurium endurance was better as the a_w of egg albumin powder reduced from 0.34 to 0.13. Cereal or grain flours are characteristically used as a component in preparation of complex foods. In general, the Food and Drug Administration (FDA) do not deem flour to be a "sensitive component" for Salmonella. Conversely, there have been situations when flour was found to be infected with Salmonella and pretreatment was performed to eradicate the pathogen, particularly when it is applied as a carrier in nutraceuticals, flavors, dried mixes in ready to cook and eat products and as boosting-caloric agent (Forghani et al. 2019). (Stojiljkovic et al. 2016) identified Salmonella enteric from pasta flour even after 1 year of storage, signifying that extended storage of pasta flour is not a proficient way for decontamination of microbial infected product. Vegetable powders, milk powders, fruit powders and dried spices, are sporadically contaminated with bacilli Salmonella. Reports on Salmonellosis related with the intake of powdered foods have been published. For instance, as stated formerly, Lehmacher et al. (1995) demonstrated a Salmonellosis connected with the digestion of potato chips fried with paprika-powdered. Though minor levels of Salmonella sustained in the product (5–50 cells/100 g), those fraction were adequate to cause foodborne disease, probably for the reason of the high fat content of the paprika-powdered potato chips. Likewise, WHO enlisted *Salmonella enteric* as second significant microorganism to be alarmed with respect to powdered infant formula PIF (Jones et al. 2019). The risks linked with PIF for the presence of *Salmonella* has been considered broadly (Beuchat et al. 2013; Oonaka et al. 2010). These works stated that there was an elevated possibility of Salmonella contamination chiefly during fortification stage and at dry mixing.

9.2.2 Cronobacter

Cronobacter sakazakii is a flagallated, non-spore forming gram negative bacteria that grow in both aerobic and anaerobic conditions within temperature range of 7–80 °C and at water activity of 0.2–0.8 for up to 12 months. This opportunistic foodborne pathogen has been studied to be competent of engendering morbidness in every age, particularly on children's leading to life intimidating hurdles (Pina-Pérez et al. 2016). *Cronobacter* has been identified in diary based food powders with infants being most infected following the intake of infested powdered infant formula (PIF). The high rates of outbreak of *Cronobacter sakazakii* in powdered infant formula demonstrate that though a great fraction of *C. sakazakii* was endowed inactivated in dried storage surroundings (Li et al. 2016), a fraction of the cells was observed greatly resistant and was alive for at least 2 years. *Cronobacter species* is renowned to have produced life-alarming infections in infants with a casualty rate of 40–80%. Contaminated food powders, particularly powdered infant formula (PIF), has been epidemiologically correlated with these foodborne illness including meningitis, seizures and necrotizing enterocolitis (Shi et al. 2017). Henceforth, World Health Organization (WHO) regarded *C. sakazakii* as one of the foodborne pathogens of main alarmed with respect to powdered infant formula (PIF).

9.2.3 Clostridium

Clostridium botulinum, a spore forming, Gram positive bacterium, is an imperative origin of food-borne illness. Various *Clostridium* species isolated are pathogenic, however regards to foodborne intoxications only two strains were found to be toxic namely *C. botulinum* and *C. perfringens* (Amuquandoh 2016). The bacilli *Clostridium* sustain on dusty surfaces and are found resistant to regular cooking and boiling temperature. *Clostridium* contamination also triggers severe condition of cramps, diarrhea and flatulence depending on whether the infection is chronic, acute, or limited. A situation of infant botulism was related owing to the ingestion of reconstituted powdered infant formula. It was described through another work, nevertheless, that sealed batches of PIF occupied in the reported case was not the origin of contamination to infants (Xin et al. 2019). Presence of spores of *C. perfringens* in the human gastrointestinal tract through contaminated food powder consumption has been studied to release enterotoxin leading to life threatening diseases. Further, *C. perfringens* apart from PIF have also been isolated from spice powders. Dried spices and herbs, involving paprika, oregano powder and black pepper contaminated with *Clostridium* species if incorporated to processed meat or vegetable dishes, give augment to infective levels of doses if the product is temperature-abused holding period. Henceforth, peculiar consideration has to be paid on spices and herb powders used in ready-to-cook and ready-to-eat foods. The microbiological safety and quality of dried herbs and spices chiefly depends upon sterile actions acquired at time of processing and during their cultivation, harvesting and post-harvest processing (Bhat et al. 1987).

9.2.4 *Bacillus*

Bacillus is a genus of Gram-positive, spore-forming and rod-shaped bacteria. Strains of *Bacillus* bacteria *B. cereus* and, very hardly, *Bacillus thuringiensis* and *Bacillus subtilis* can generate toxins that are harmful to humans. Thermostable enterotoxin (haemolysin) is released by *Bacillus* species present in food products when exposed to gastrointestinal tract (Kim et al. 2015). *Bacillus cereus* spores have been reported to sustain in desiccated and dehydrated foods such as rice flour, cereal powder, and in dry food handling surroundings for lengthy time periods, and can evolve and develop in recombined (reconstituted) food powders to great extent when not suitably processed or stored at appropriate temperatures (Haughton et al. 2010). Spores of this strain is to enormous range is capable to generate spores that can sustain pasteurization, and endure the preparation process of powdered foods peculiarly powdered milk as most *B.cereus* strains isolated from dairy powder products are capable of emerging, and releasing toxic compounds below 10 °C to temperature condition high as 50 °C (Bursová et al. 2018). Although epidemiological verification relating to *B.cereus* through PIF is rare, diarrhoea is a noteworthy reason of ill health and fatality in children particularly new born in developed countries.

9.3 Issue of Existing Decontamination Techniques on Achieving Microbial Safety of Food Powders

Commonly used techniques of microbial decontamination for food powders were introduced on the application of temperature, with varying temperature limits and processing time. The conventional techniques say canning, sterilization, pasteurization and steaming under elevated conditions of temperature applied till date have assured microbiological quality of powdered foods by gaining targeted values of 2–3 log reductions although 5–6 log reductions needs to be achieved according to Code of Federal Regulation for every food product to be microbially protected (Dufort et al. 2017). Pasteurization was found to destroy a considerable fraction of the microbial count. Nevertheless, samples that endure this procedure need particular storage environment. In fact, pasteurized food powders yet holding microbes have been reported to be capable of multiplying under favorable conditions making the product unsuitable for human intake. Alike, sterilization has also been applied for an extended period to decontaminate food pathogens. Unfortunately the application of sudden heat shock could lead to enhanced resistance to consequent decontamination process as observed in case of *Cronobacter sakazakii* and *Listeria monocytogenes* (Lin and Chou 2004). However, the detrimental effect on sensorial and nutritional properties that emerges on the use of elevated temperature is the cause which confines the application of direct heat processing regardless of the absolute decontamination levels. Moreover, the use of thermal techniques in food processing industries stands inadequate due to its restricted heat transfer efficiency and low final product quality and uneven product heating. Further, unlike fluid foods microorganism decontamination in food powders falling to solid state arises major hurdle; say difficulty in penetration of heat waves to powder sample owing to the bulk nature of powder samples.

Hence, there was a requisite for the use of novel decontamination techniques in powdered foods to attain micobially safe powders with high-quality. Hence, understanding the growing concern of producing a superior-quality powdered foods with high microbiological safety with enhanced functionalities, novel techniques particularly non-thermal process established its relevance in powdered food decontamination (Zhang et al. 2006). The comprehensively assuring novel techniques for food powder decontamination marks to be pulsed electric field (PEF), ozone processing (OP), non-thermal plasma (NTP) and high hydrostatic pressure processing (HHP) among which majority is been recognized, and few say HPP and PEF are at the threshold of reorganization.

9.4 Novel Technologies to Assure Microbiological Safety in Food Powders

9.4.1 Pulsed Electric Field Processing (PEF)

PEF is the most widely applied non-thermal technique which has been used for microorganism decontamination with safeguarding end product quality of food products. PEF comprises the method of exposing the food products placed among two electrodes by short pulses and large voltages, with electropermeabilization being the main principle. Processing chamber, power generator, pulse supply, and monitoring systems are the major parts of the PEF unit (Barbosa-Cánovas et al.

1999). PEF method has achieved much attention as it deploys a result analogous to that of conventional process but does not alter the structural and organoleptic property parameters of the sample. Broad microbial decontamination methods have been conducted to confirm the notion of pulsed electric field as a green food pasteurization process. The main effect of pulsed electric field process is electroporation of the cell membrane of microorganism, that is, improved outer wall permeability as an outcome of enlargement within fissures in the cell wall under the circumstance of a field created adjoining the cell membrane. Application of very low pulses for few seconds at different voltage levels have been studied in a small number of powdered foods for microbial decontamination without destructing the quality grade of the powdered foods (Roohinejad et al. 2018). Further as cell breakage mainly relies on the moisture content available within the food, powdered foods that belongs to low water activity food group was studied to be little suitable for microbial inactivation to considerable levels via pulsed electric field technique. Various authors suggested that microbial inactivation of about utmost of 4 log reduction was attained using pulsed electric field process on powdered foods say, corn meal, pea powder, spice powder, and dill powder (Pina-Pérez et al. 2016). Further authors suggested that though PEF processing retained the whole powder properties after treatment it was not recommended as a potential process to inactivate microbial count in food powders comparison to other novel decontamination process with respect to log reduction achieved.

9.4.1.1 PEF Application for Microbiological Safety of Food Powders

The initial research on microbial decontamination employing pulsed electric field processing was to evaluate the impact of pulsed electric field on *Bacillus subtilis* and *E.coli* on ready to cook soup mix powder (Vega-Mercado et al. 1996). A reduction of 5.3D and 6.5D was achieved at pulse rate of 30 pulses at 33 kV/cm for targeted bacilli species respectively. The decontamination rate of microbial reduction in the above study was studied to resemble thermal heating and ultraviolet technique. Soon, another researcher (Keith et al. 1997) studied the result of pulsed electric field processing for log reduction of few bacilli and fungi. The plate count of the ordinary spice powders was in the array 3.0×10^5 to 1.2×10^6 CFU/g for sweet basil and 6.1×10^4 to 2.5×10^5 CFU/g for aromatic vegetable powder. At 40 kV/cm the highest inactivation range of 1 log reduction was demonstrated for the chosen species of microorganisms in sweet basil and aromatic vegetable powder sample correspondingly preserving nutrients, taste and colour of both powder samples. In above study gompertz growth kinetics model was applied to assess the decontamination kinetics. Later, experiments was performed to investigate the effect of pulsed electric field technique for C. sakazakii inactivation in cocoa fortified PIF (Pina-Pérez et al. 2013a, b). Results suggested that suggested decontamination array of 4.4 $\log_{10}$ cycles, was achieved during a PEF treatment with 3000 μs at 15 kV/cm. The endurance plot for *C. sakazakii* in the aforementioned work was determined applying biphasic model. Further, contrasting extensive application of pulsed electric field for microorganism decontamination in fluid food products, its relevance for powdered food decontamination is little restricted owing to the microbial confrontation to inactivation was probably coupled to the happening of bacilli spores.

9.4.2 High-Hydrostatic Pressure Processing (HHP)

High-hydrostatic pressure processing (HHP) commonly called as high pressure processing technique, is a novel non-thermal process applied for food preservation as well as for product preperation. HHP process includes the technique of holding powdered foods in or out containers to pressure range that usually may vary from 100–800 MPa. HHP in food applications is controlled by two relevant laws: primarily, the isostatic principle and secondly, Le chatelier's principle (Oey et al. 2008). HHP has established a wide usage in food processing sector due to its nominal heat generation, product uniformity, improved product shelf life and least amount energy necessity. The prior relevance of HHP was demonstrated for pasteurization of milk. In a while scientist group was in an urge to discover the impact of HHP for microbial decontamination in solid foods. Cases of HHP subjected products commercialized in the retail involve meat, horticultural and dairy products (Yuste et al. 2001). Captivatingly, the control of high-hydrostatic pressure technique for microbial inactivation of powdered foods has too been demonstrated. The key mechanism of HHP technique is demolition generated to the non-covalent molecules existing inside microorganisms say nucleic acid, proteins and fats, leading to damage of the cellular membrane and progressively ensuing in microbial destruction. The mechanism followed for microbial decontamination in HHP is as explained. Revelation to very large pressure levels for a portion of little seconds ceases cellular wall cleavage and a decline of intercellular acidic value within the cell of microorganism. It is also been found that swift volumetric expansion outcomes in a structural variations of major cellular compounds say proteins, thus stimulating unfolding and denaturation of protein molecules (Domitrovic et al. 2006). Further, minor compounds that include amino acids, minerals, vitamins and compounds aiding to the quality aspect of food powder was observed to sustain not affected. However dissimilar to processing condition for fluid foods the robust reliance of microbial decontamination on water activity was found as the key aspect that highlights the decontamination kinetic in powdered foods during HHP application.

9.4.2.1 HHP Application for Microbiological Safety of Food Powders

A technique proficient of decontaminating powdered foods at a small cost using HHP was developed by Tsujimoto et al. (2004). The fabricated HHP system included a roller compactor operating on shear principle, making it appropriate for microorganism decontamination of food powders. The work was conducted on few powdered food samples say: fennel powder, corn flour and herb powders contaminated with *B. cereus*. Later, (Marco et al. 2011) described decontamination of *Bacillus cereus* cells in food powder media using application of HHP. Statistically higher levels of decontamination were observed at HPP pressure of 200 and 500 MPa for treatment time of 10 min. Various studies have discovered the prospective application of HHP process to decontaminate *C. sakazakii* cells in PIF. The foremost study on *Cronobacter sakazakii* inactivation using HHP has been conducted by (González et al. 2006). A decontamination values of two, four, six $\log_{10}$ cycles was acheived. Later, (Arroyo et al. 2012) confirmed that reduced HHP treatment resulted in inactivation range of *C. sakazakii* cells in PIF in range 5–7 $\log_{10}$ cycles. However, in spite of the pros that HHP has on decontaminating microbial count to targeted values, there are few demerits that could vaguely change the food powder properties and consequently the food quality. As studied with Windyga et al. (2008) on microbiological decontamination of coriander and caraway powder for 30 min HHP exposure at 800 and 1000 MPa under helium atmosphere. The authors demonstrated that HHP under helium atmosphere, under adequate time and pressure enhanced the microbiological value of spice powders. On the other hand, the robust reliance of microbial decontamination on a_w is even now a significant hindrance to the application of HHP as a decontamination technique in herbs and food powder manufacture. Therefore, to maintain our powder food value to convene quality of export worth it is important to control the processing time and pressure during HHP for which further studies are necessary to be conducted.

9.4.3 Pulsed Light

Non-thermal PL technology is a novel technique, where short period pulses of a high charge flash, surplus in UV radiations are frequently and swiftly exposed to the powdered surface of food by using a suitable lamp source. PL produces a broad spectrum of white beam with maximum emanation between 300 and 500 nm. The microbial decontamination impacts of PL wavelength due to the assimilation of the voltage by robustly conjugated bonds between carbon molecules in proteins of microorganisms, leading to breakage in cell metabolism was revealed primarily (Kao et al. 2005). The mechanism of PL technique to assure microbial safety has been extensively studied. The destruction of DNA in the microbes assisted by damage to cell configuration have been regarded as the chief aspect for microbial lethal result. The devastation of microbes ceases with the genetic molecule (DNA/RNA) engrossing emitted UV radiation by PL, proceeded by the development of cross linked molecules, resulting a mutation in the genetic material of targeted microbial cell. This process accordingly stops DNA functioning and affects further replication of the microbial cells. From then, fascinatingly, the impact of PL on decreasing microbial load within fluid and food powders has been performed. (Koch et al. 2019). Nevertheless, it can be concluded that out of many parameters that determines efficacy of PL technique, flash frequency and energy level marks to be the principal factors in attaining preferred microbiological safety levels respect to food powders.

9.4.3.1 PL Application for Microbiological Safety of Food Powders

In respect to application of PL technique for microbial decontamination of powdered foods, wheat flour and pepper powder has been researched for inactivation of *S. cerevisiae* by (Fine and Gervais 2004). At 31.12 J/cm^2, a diminution of 2.93 and 0.7 was obtained for wheat flour and pepper powder respectively. The growth curve was considered applying Baranayi logistic model. In spite, it was studied that colour and flavor properties of the studied powdered foods were significantly maintained at the set processing parameters. Impact of PL technique on corn flour has also been studied. At a pulse rate of 100 s pulses at distance of 3 cm (Jun et al. 2003) attained 4.93 log reduction in *A. niger* spore contaminated in corn flour. Through another research, the decontamination efficacy of PL process for *L. monocytogenes* isolated from PIF was studied (Choi et al. 2010). In the course of this work it was discovered that commencing from initial load of 10^{5}CFU/g, approximate to four-five log reduction of the microbial cells was achieved with PL technique at various pulse rate and voltage. Authors reported that greatest inactivation of targeted cells was obtained at 25 kV. In addition, (Nicorescu et al. 2013) observed a significant reduction in the *Bacillus subtilis* cells at PL treatment (10 J/cm^2) to 0.8 log reduction in pepper powder and caraway powder. Prior works conducted for PL technique have exposed that PL had negligible effect on the nutritional quality and organoleptic properties of treated powders, thus assuring it to be a secure technique giving desired quality and secure final powder product.

9.4.4 Ozone Processing (OP)

Ozone (O_3) is an inorganic molecule that exists in a gaseous phase and is formed from the molecule of oxygen due to the mutual effect of ultraviolet rays and electric discharges in the atmosphere. The chief possible advantage of ozone technique is that excess ozone self-degrades to generate oxygen, leaving zero remnants. The great significance in ozone as an decontaminating agent is regards to its wide anti-microbial spectrum (Garud et al. 2019). The microbial inactivation initiates as the bacterial cell riveting released ozone, and the inactivation inactivation is chiefly accredited to anti-microbial nature of ozone molecule. The cellular constituents present in the microbial cell oxidizes, due to the increased oxidation potency of ozone. As oxidized ozone molecule begins to enter the cell of a besieged organism, all vital cell constituents including, DNA, proteins and enzymes go through total oxidation, leading to entire cell rupture (Brodowska et al. 2017). As a strong disinfectant, it has marked broad range relevance in the food processing industry. Further, as per the current FDA endorsement of ozone gas as a secure preservative, efficacy of ozone molecule in food relevance augmented, peculiarly in liquid foods. However, the impact of ozone technique on decontamination of powdered foods have also been studied proficiently. However through framing of this chapter it was considered that produced ozone molecule and exposure stage of powdered foods to ozone technique stood to be the key factors to accomplish preferred point of microbial log reduction respect to powdered foods.

9.4.4.1 Ozone Processing Application for Microbiological Safety of Food Powders

Microbiological safety levels attained on paprika powder on revelation to ozone processing was studied (Akbas and Ozdemir 2008). Ozone levels of 1, 5, 7 and 9 ppm for 6 h were released on *B. cereus* cells, whereas 0.1, 0.5 and 1.0 ppm were used on *E. coli* cells. Results discovered that contact to ozone levels of 5 ppm and 1 ppm for 6 h was discovered as the best grouping in attaining highest decontamination of *Bacillus cereus* and *E. coli* respectively, preserving powder particle properties significantly. In before time, Byun et al. (1998) demonstrated the relative impact of irradiation and ozone processing on microbial quality of ginseng powder. Ozone processing at concentration of 18 ppm c for 8 h and irradiation at 7.5 kGy were carried out to calculate approximately its decontamination efficacy on yeast and total bacteria with count of 6.7 × 102 CFU/g and 1.8 × 105 CFU/g. It was observed that o ozone treatment resulted in 3 log reduction approximately of targeted microbes. A sudden deprivation of (2 log reduction) of mycotoxin generated by Fusarium in rice flour was demonstrated by (Young et al. 2006). It was found that as much as ozonation of powdered food is regarded, mycotoxin possessed a quick deprivation at low pH 4–6 compared to pH of 7–8. Likewise, (Tiwari et al. 2010) described that an ozone exposure of 0.16 mg/gm barley powder led in attaining 96% destruction of spore of fungi within 5 min of ozone exposure. In general it can be stated that studies performed on powdered foods discovered that, when dosage of ozone, its exposure time was maintained within the critical value, it maintained the powder properties such as color, odour, and Vitamin C enhancing the consumer reception worth of the product and ozone technique for food powder decontamination.

9.4.5 Non-thermal Plasma (NTP)

Plasma, the ionized state, describes wholly and partly ionized gas constituted mainly of atoms, ions and other components creating plasma to be neutrally charged molecule (Bourke et al. 2018). Food sample to be decontaminated is treated openly or in some way to the plasma molecules ionized using an exterior energy resource or using electric field. The accurate mechanism of plasma technique for decontaminating microbial cells was considered to be on the basis of elementary progression of, erosion of microbial cell throughout photo-desorption (Fernández and Thompson 2012). Currently, NTP technique has been originated to be applied enormously in food industries for food surface modification, inactivation of undesired compounds and volatile oil extraction. Nevertheless, lately, the progress in studies has revealed the vast possibility of non thermal plasma technique as a non-thermal inactivation technique producing a deadly consequence on an extensive series of microbes in fluid as well as powder foods.

9.4.6 *Non thermal Processing Application for Microbiological Safety of Food Powders*

The practicability of microwave plasma exposure to infested microbial cells sprayed onto, pepper powder, oregano and red chili powder samples was studied by Hertwig et al. (2015a, b). It was recognized that microwave plasma treatment reduced the microbial count of the pepper, paprika and oregano powder by more than 3 log reductions respectively. The decreased decontamination level of bacterial spore in oregano powder was owing to the reduced primary microbial count. Through another research, DBD plasma technology on grouping with radiofrequency thermal technique at 1500 and 1000 W has been considered effectual in dropping the bacterial cell load to one log CFU/g in black pepper (Choi et al. 2018). It has been confirmed that synergistic effect of DBD plasma technology and radiofrequency thermal technique resulted in a insignificant transformation in color, moisture content and a_w.

The results revealed that cold plasma treatment at 1000 W for 1.5 min eliminated the development of desire fungal cells by 50%.

In general, powders' being dispersive and fine in nature with great potential to acquire moisture on revelation to environmental conditions, the proper packaging and storage is the critical processing operation to be carried out. In-package sanitization by any of the aforementioned novel non-thermal technique has massive potential and attained ever-increasing interest from the food powder industry recently regards to its proficacy in microbial safety and quality (Patil et al. 2014). A summary of the salient features of works on microbial diminution levels attained for food powders through revelation to novel decontamination techniques (i.e. non-thermal treatments) are described in Table 9.1.

9.5 Future Trends to Assure Microbial Safety in Food Powder Industry: An Outlook

Though contaminated poultry, as well as horticultural and dairy produce, by food-borne microbes will persist to pretense considerable food security problems, the time ahead might reveal advance issues related with low water activity, dried and powdered foods. It is believable that powdered food products very rarely considered at risk for food related illnesses may come into view as issue of major concern. *Salmonella* will likely continue to pose the greatest threat in these foods, due to its

Table 9.1 Summary of microbial diminution levels attained for food powders on exposure to non-thermal treatment applications

Technology	Microbial targets	Powder type	Treatment condition	Salient outcomes	Reference
HPP	*Bacillus cereus*	Corn flour, fennel powder & Chinese herbs	Pressure of 6.8×10^4N/cm	Spore colony count decreased to significant level	Tsujimoto et al. (2004)
	Bacterial strains	Licorice powder	HPP at 0.06 MPa	Attained 5 log reduction	Marco et al. (2011)
	Bacillus cereus	Olive powder	400 MPa for treatment period from 10 to 225 min	Improved spore inactivation by 0.5 log cycles	Black et al. (2008)
	Cronobacter sakazakii	Infant formula	Exposed to pressure of 200, 400 & 500 MPa	Achieved decontamination level of 2, 4, 6 $\log_{10}$ cycles	Pina-Pérez et al. (2011)
	Cronobacter sakazakii	Infant formula	Treated at 100–200 MPa for 10–20 min	Inactivation count of 7 $\log_{10}$ cycles attained	Arroyo et al. (2012)
	Cronobacter sakazakii	Infant formula	Exposed to pressure of 400 & 600 MPa	*Cronobacter spp*. inactivation obtained to significant levels	Pina-Pérez et al. (2012)
	Total aerobic bacteria and yeasts and mold	Garlic powder	HPP at 600 MPa for 5 min	Decreased total aerobes count and yeasts count by 1.62 and 1.43 log CFU/g	Park et al. (2018)
	Naturally occurring microorganisms	Red bean powder	Exposure to HPP at 400, 500, and 600 MPa for 5 min	Microbial count reduced to 1.83, 1.55, and 1.05 log CFU/g	Lee et al. (2018a, b)

Table 9.1 (continued)

Technology	Microbial targets	Powder type	Treatment condition	Salient outcomes	Reference
PEF	*Escherichia coli* & *Bacillus subtilis*	Pea soup mix powder	PEF voltage of 33 kV/cm and 30 pulses	5.3D reduction was observed	Vega-Mercado et al. (1996)
	Total coliform, *E. coli*, yeast and mold	Onion, basil and dill powder	Treatment performed at PEF voltage of 26, 40 and 28 kV/cm	One-log reduction was attained	Keith et al. (1997)
	Cronobacter sakazakii	Infant milk formula	Low intensity PEF treatment of 15 kV/cm applied for 3000 μs	Decontamination levels of 4.4 $\log_{10}$ cycles was achieved	Pina-Pérez et al. (2013a, b)
	Aerobe, yeast/mold, and coliform	Tangor juice powder	PEF processing at 70 °C–16 kV/cm–100 kJ/L	The aerobe, yeast/mold, and coliform counts decreased by 3.9, 4.3, and 0.8 log CFU/mL	Lee et al. (2018a, b)
	E. coli	Stevia & Ginseng powder	20 kV/cm, 270 μs	15.11 ± 0.11 mm was greatest inhibition zone diameter obtained	Pina-Pérez et al. (2018)
	Native microorgamisms	Dried blueberries	PEF voltage of 2 kV/cm	Log reduction (5D) attained	Yu et al. (2016)

Technology	Microbial targets	Powder type	Treatment condition	Salient outcomes	Reference
PL	*Scharomyces cerevisiae*	Black pepper & wheat flour	Treated at energy level of 31.12 J/cm^2	Log reduction of 2.93 and 0.7 was obtained	Fine and Gervais (2004)
	Aspergillus Niger	Corn meal	Treated at energy level of 5.6 J/cm^2	4.93 log reduction was achieved	Jun et al. (2003)
	Listeria monocytogenes	Powdered infant formula	PL treatment for 5000, 600, 300 and 100 μs at 25 kV	4–5 log reduction of microbial count was attained under set conditions	Choi et al. (2010)
	Bacillus subtilis	Ground caraway, black pepper powder	Exposed to PL of 10 J/cm^2	1 & 0.8 log reduction in pepper and caraway was attained	Nicorescu et al. (2013)
	Bacillus subtilis	Peppercorn powder	Treated at at intensities of 4 and 6 pulses respectively	A decimal reduction of 2.1 & 2.6 log CFU was achieved	Moreau et al. (2009)
	Salmonella typhimurium & *Escherichia coli*	Red pepper	Sample exposed to light intensity of 20.4 kJ/m^2 for 10 min	Obtained 0.22 CFU/g log reduction	Cheon et al. (2015)
	Cronobacter sakazakii	Powdered infant formula	Powder exposed to 57.5 ± 0.7 °C for 28 seconds	Exhibited maximum inactivation of 3.18 $\log_{10}$ CFU/g	Chen et al. (2018)
	Bacterial load	Dried sesame seeds	PL treatment at a total fluence of 39.85 J/cm^2	0.86 log microbial reduction was attained and at 44.46 J/cm^2 reduction enhanced to 1.02 log reduction	Hwang et al. (2017)
	Listeria species	Powdered infant formula	PL treatment at a total fluence of 17 mJ/cm^2	Inactivated >99% of the vegetative cell populations Listeria species	Arroyo et al. (2017)
	Salmonella Enteritidis	Dried almonds	Sample exposed to pulsed light at 3000, 3400 and 3800 V at distances of 14.1 or 19.1 cm for 20 or 60 s	PL decreased *Salmonella* count by 0.44 to 4.14 log CFU	Eser Öner (2017)

(continued)

Table 9.1 (continued)

Technology	Microbial targets	Powder type	Treatment condition	Salient outcomes	Reference
OP	*E. coli* & *Bacillus cereus*	Flaked red pepper	O_3 dose of 1 ppm and 5 ppm for 360 min	Attained maximum inactivation of 5 log reduction	Akbas and Ozdemir (2008)
	Total aerobic bacteria & yeast	Red ginseng powder	Exposed to 18 ppm ozone concentration for 8 h	Attained up to 2 log reduction	Byun et al. (1998)
	Fusarium	Rice flour	Exposed to 10 ppm ozone concentration for 9 h	2 log reduction was achieved	Young et al. (2006)
	Fungus	Barley powder	Ozone dose of 0.16 mg/gm was applied	Obtained 96% spore inactivation in 5 min	Tiwari et al. (2010)
	Aspergillus flavus	Peanut flour	Ozone dose of 5 ppm with treatment time (15–10 min) was applied	Achieved a inactivation of 80% and 77%	Proctor et al. (2004)
	Enterobacteriacea	Oregano, thyme, mountain tea, lemon verbena and chamomile powder	Exposed to ozone treatment of 4 ppm for 30 or 60 min	Exhibited a 4 log reduction, from around 6.5 to 2.5 in the case of oregano, and only a 1–2 log reduction for other four herb powders	Kazi et al. (2017)

Technology	Microbial targets	Powder type	Treatment condition	Salient outcomes	Reference
NTP	Bacterial flora	Paprika, pepper &oregano powder	The frequency of the microwave plasma was2.45 GHz and was ignited for 7 s	Bacilli of paprika, pepper & oregano by more than $3\log_{10}$, $3\log_{10}$ & $1.6\log_{10}$ respectively	Hertwig et al. (2015a)
	E. coli	Pepper powder	DBD-PT & RF-TT at 1000 W and 1500 W	Effectively inactivated *E. coli* to 9 log CFU/g	Choi et al. (2018)
	Aspergillus flavus	Red pepper	NTP at 900 W for 20 min	Achieved 2.5 ± 0.3 log spores/g reduction	Kim et al. (2014)
	Bacillus cereus spore, *Escherichia coli* & *Aspergillus brasiliensis*	Onion powder	MCPT at 400 W for 40 min	Effectively decreased the count of *A. brasiliensis*, by 1.6 log spore/cm^2	Kim et al. (2017)
	Aspergillus flavus & *Bacillus cereus*	Red pepper	MCPT at 0.17 W/m^2 and 0.15 W/m^2	Spore count reduced by0.7 ± 0.1 & 1.5 ± 0.2 log spores/cm^2.	Kim et al. (2018)
	Fusarium oxysporum	Paprika powder	NTP at 1000 W for 90 seconds	Eliminated the fungal growth by 50%.	Go et al. (2019)
	Listeria monocytogenes, E. coli and Salmonella enterica	Onion powder	Exposure to 15 kHz helium NTP	Resulted in 3.1 log CFU/cm^2 reduction	Kim and Min (2018)
	Bacillus subtilis, E. coli, Salmonella Enteritidis	Black pepper powder	Plasma was generated at atmospheric pressure in ambient air for 300 s	Decreased *B. subtilis* from 7.36 to 2.30 CFU/g, *E. coli* and *Salmonella Enteritidis* to below detection level	Mošovská et al. (2018)
	Bacillus cereus & *Aspergillus flavus*	Red pepper flakes	MCPT at 900 W for 20 min	*B. cereus* and *A. flavus* spores were reduced by 1.4 ± 0.3 & 1.5 ± 0.2 log spores cm^{-2}	Kim et al. (2018)
	Aspergillus flavus, Alternaria alternata and *Fusarium culmorum*	Dried maize seeds	Exposed to DCSBD of power density 80 W cm^{-3}	A reduction of 3.79 log (CFU/g) in *F. culmorum*, 4.21 log (CFU/g) in *A. flavus* and 3.22 log (CFU/g) in *A. alternata*	Zahoranová et al. (2018)
	Fusarium oxysporum	Paprika powder	Exposed to NTP at 1000 W for 90 seconds	Controlled the fungal growth by 50%.	Go et al. (2019)
	E. coli & *Staphylococcus aureus*	Herbal tea powder	Exposed to electrical discharge plasma of 10 kHz and power of 80 W	-values of 0.73 and 0.67 min were obtained	Chingsungnoen et al. (2018)

uncanny ability to survive desiccation. The use of novel processing technologies was confirmed to embrace elevated prospective for manufacturing powdered foods of significant quality. During past decade, various studies have established the decontamination effect of PEF, PL and OP in a variety of powdered foods say onion powder, soup mix powder, PIF, basil powder, spice powders and cereal flours. Of aforesaid three methods, first two were observed to be broadly used for decontamination of bacilli spores (*Salmonella* and Cronobacter) present regularly in PIF and herb powders while, latter was observed capable in resolving the highly resistant spores *(A. niger, Fusarium)* present in vegetable and spice powders (Pina-Pérez et al. 2013a, b). Further, in contrast, disclosure to HHP though inhibited development of critical microbes to considerable levels was established to have an effect on the dietary quality, micromolecular structures and property within the food powders. Among the novel technologies overviewed, non-thermal plasma stands to be the major competitive respect to the swift microbial cell inactivation. This is due to the hurdle effect of exposure to UV, generation of free radicals and reactive species during plasma generation. Further, the produced electric field in direct manner results in absolute inactivation of microbes to significant levels in powdered foods.

Nevertheless, all these novel techniques are yet in an infant phase of development regarding their relevance in the food powder manufacturing while assuring the safety and quality respect to microbiological aspects. The most eminent techniques regards to overall powdered food microbial quality, safety and practical point of view, can be categorized as follows: NTP > PL > HHP > PEF > OP. To validate the suitability of product and technique, the subsequent pace could be to level up the desired technique to industrial level. Within upcoming years HHP and NTP would be commercialized in majority of the countries across the globe (Morales-de la Peña et al. 2010). In contrast, most novel non-thermal processing techniques have not been commercialized for food powders still. Henceforth, the use of these novel technologies to inactivate microbial load in food powder still creates extra studies to attain industrial scale for straight powder decontamination. The existing restriction of these methods in the industrialized level is associated to elevated primary investment expenditure and deficit of data concerning process control. To eliminate these troubles, it is imperative to optimize the processing condition and consider the usefulness of the treatment, dependent parameteres, matrix medium and initial microbial load. Regards to the food powder, development of proficient treatment technique competent of retaining powder nutrition characteristics and microbial safety due to elevated temperature coverage are the important contemplation respect to the industry. Therefore it can be summarized that, the application of two or more nonthermal technologies together or simultaneously materialize to produce enhanced treatment efficacy. However, a systematic estimation of costs related with launching, working and maintaining the flow line and product line considering the practical application in industrial level needs to be calculated to manufacture food powders of superior microbial safety with nutritional qualities and organoleptic properties at desired level as demanded by consumers and food industries.

9.6 Conclusion

Food powders or dried foods (aw <0.70) were formerly presumed to be microbiologically safe. Conversely, they were observed to be contaminated with food-borne illness causing microbes, and established to be the source of various food-borne outbreaks. This chapter will present a thought initiation on the microbiological challenges regards to safety of food powders, including milk powder, PIF, cereal flours, spices, fruit and vegetable powders particularly in light of current microbial breakouts and recalls, with providing insights on novel techniques that was found proficient for microbial decontamination of food powders following regulatory standards.

Regions of alarm comprise the secure sourcing of raw materials used in powder processing, unsanitary drying practices, insanitary storage conditions, faulty equipment design, uncontrolled food powder reconstitution process, cross-contamination of food powders, unacceptable testing techniques, unsuitable sampling plans, malfunction to act on discovered food-borne microbe-positive samples, failure to execute an standard HACCP guidelines in processing line, improper importing or exporting criteria The major outbreak of powdered food was observed through spice and dairy powders which sickens more than thousand people causing death when a outbreak occurs. However, regards to the published literature till date, the most probable food-borne pathogens of disquiet from food powder will continue to be *Salmonella* and *Cronobacter*.

It is expected that newly formulated food powder products may be less considered at risk for contamination in the future owing to the adaptation of novel thermal techniques discussed in previous sections and execution of good manufacturing practices and hazard analysis points in the powder processing line. Various novel techniques, for instance, NTP, PL, PEFand HHP have been broadly considered for inactivation of *Salmonella*, *Bacillus, Cronobacter*, and spoilage yeast. Furthermore, hurdle effects of novel technology combinations with implementation of food safety standards may be established to be extremely potential for tackling the most challenging cells and species in existing food powders.

Conclusively, to assure that food powders are entirely microbiologically safe use of a novel non-thermal processing during its processing or post-processing with quality control standards followed is to be assured. In reverse case, Powdered foods can no more be regarded that they are intrinsically protected from food borne pathogens just owing to the fact that they are dried, preferably, an –"if…"sentence will for all time be proceeded.

References

Abatcha, M. G., Zakaria, Z., Kaur, D., & Thong, K. L. (2014). Occurrence of antibiotic resistant Salmonella isolated from dogs in Klang Valley, Malaysia. *Malaysian Journal of Microbiology*. https://doi.org/10.21161/mjm.58213.

Akbas, M. Y., & Ozdemir, M. (2008). Effect of gaseous ozone on microbial inactivation and sensory of flaked red peppers. *International Journal of Food Science & Technology, 43*(9), 1657–1662. https://doi.org/10.1111/j.1365-2621.2008.01722.x.

Amuquandoh, F. E. (2016). *Essentials of food safety in the hospitality industry*. Bloomington: Xlibris.

Arroyo, C., Cebrián, G., Condón, S., & Pagán, R. (2012). Development of resistance in Cronobacter sakazakii ATCC 29544 to thermal and nonthermal processes after exposure to stressing environmental conditions. *Journal of Applied Microbiology, 112*(3), 561–570. https://doi.org/10.1111/j.1365-2672.2011.05218.x.

Arroyo, C., Dorozko, A., Gaston, E., O'Sullivan, M., Whyte, P., & Lyng, J. G. (2017). Light based technologies for microbial inactivation of liquids, bead surfaces and powdered infant formula. *Food Microbiology, 67*, 49–57. https://doi.org/10.1016/j.fm.2017.06.001.

Barbosa-Cánovas, G. V., Góngora-Nieto, M. M., Pothakamury, U. R., & Swanson, B. G. (1999). Fundamentals of High-Intensity Pulsed Electric Fields (PEF). In *Preservation of foods with pulsed electric fields* (pp. 1–19). https://doi.org/10.1016/b978-012078149-2/50002-7.

Beuchat, L. R., Komitopoulou, E., Beckers, H., Betts, R. O. Y. P., Bourdichon, F., Fanning, S., et al. (2013). Low–water activity foods: Increased concern as vehicles of foodborne pathogens. *Journal of Food Protection, 76*(1), 150–172. https://doi.org/10.4315/0362-028x.jfp-12-211.

Bhat, R., Geeta, H., & Kulkarni, P. R. (1987). Microbial profile of cumin seeds and chili powder sold in retail shops in the City of Bombay. *Journal of Food Protection, 50*(5), 418–419. https://doi.org/10.4315/0362-028x-50.5.418.

Black, E. P., Linton, M., McCall, R. D., Curran, W., Fitzgerald, G. F., Kelly, A. L., & Patterson, M. F. (2008). The combined effects of high pressure and nisin on germination and inactivation of Bacillusspores in milk. *Journal of Applied Microbiology, 105*(1), 78–87. https://doi.org/10.1111/j.1365-2672.2007.03722.x.

Blessington, T., Theofel, C. G., & Harris, L. J. (2013). A dry-inoculation method for nut kernels. *Food Microbiology, 33*(2), 292–297. https://doi.org/10.1016/j.fm.2012.09.009.

Bourke, P., Ziuzina, D., Boehm, D., Cullen, P. J., & Keener, K. (2018). The potential of cold plasma for safe and sustainable food production. *Trends in Biotechnology, 36*(6), 615–626. https://doi.org/10.1016/j.tibtech.2017.11.001.

Brodowska, A. J., Nowak, A., & Śmigielski, K. (2017). Ozone in the food industry: Principles of ozone treatment, mechanisms of action, and applications: An overview. *Critical Reviews in Food Science and Nutrition, 58*(13), 2176–2201. https://doi.org/10.1080/10408398.2017.1308313.

Bursová, Š., Necidová, L., & Haruštiaková, D. (2018). Growth and toxin production of Bacillus cereus strains in reconstituted initial infant milk formula. *Food Control, 93*, 334–343. https://doi.org/10.1016/j.foodcont.2017.05.006.

Byun, M.-W., Yook, H.-S., Kang, I.-J., Chung, C.-K., Kwon, J.-H., & Choi, K.-J. (1998). Comparative effects of gamma irradiation and ozone treatment on hygienic quality of Korean red ginseng powder. *Radiation Physics and Chemistry, 52*(1–6), 95–99. https://doi.org/10.1016/s0969-806x(98)00082-6.

Chen, D., Wiertzema, J., Peng, P., Cheng, Y., Liu, J., Mao, Q., et al. (2018). Effects of intense pulsed light on Cronobacter sakazakii inoculated in non-fat dry milk. *Journal of Food Engineering, 238*, 178–187. https://doi.org/10.1016/j.jfoodeng.2018.06.022.

Cheon, H.-L., Shin, J.-Y., Park, K.-H., Chung, M.-S., & Kang, D.-H. (2015). Inactivation of foodborne pathogens in powdered red pepper (Capsicum annuum L.) using combined UV-C irradiation and mild heat treatment. *Food Control, 50*, 441–445. https://doi.org/10.1016/j.foodcont.2014.08.025.

Chingsungnoen, A., Maneerat, S., Chunpeng, P., Poolcharuansin, P., & Nam-Matra, R. (2018). Antimicrobial treatment of Escherichia coli and Staphylococcus aureus in herbal tea using low-temperature plasma. *Journal of Food Protection, 81*(9), 1503–1507. https://doi.org/10.4315/0362-028x.jfp-18-062.

Chitrakar, B., Zhang, M., & Adhikari, B. (2018). Dehydrated foods: Are they microbiologically safe? *Critical Reviews in Food Science and Nutrition, 59*(17), 2734–2745. https://doi.org/10.1080/10408398.2018.1466265.

Choi, E. J., Yang, H. S., Park, H. W., & Chun, H. H. (2018). Inactivation of Escherichia coli O157:H7 and Staphylococcus aureus in red pepper powder using a combination of radio frequency thermal and indirect dielectric barrier discharge plasma non-thermal treatments. *LWT, 93*, 477–484. https://doi.org/10.1016/j.lwt.2018.03.081.

Choi, M.-S., Cheigh, C.-I., Jeong, E.-A., Shin, J.-K., & Chung, M.-S. (2010). Nonthermal sterilization of Listeria monocytogenes in infant foods by intense pulsed-light treatment. *Journal of Food Engineering, 97*(4), 504–509. https://doi.org/10.1016/j.jfoodeng.2009.11.008.

Domitrovic, T., Fernandes, C. M., Boy-Marcotte, E., & Kurtenbach, E. (2006). High hydrostatic pressure activates gene expression through Msn2/4 stress transcription factors which are involved in the acquired tolerance by mild pressure precondition inSaccharomyces cerevisiae. *FEBS Letters, 580*(26), 6033–6038. https://doi.org/10.1016/j.febslet.2006.10.007.

Dufort, E. L., Etzel, M. R., & Ingham, B. H. (2017). Thermal processing parameters to ensure a 5-log reduction of Escherichia coli O157:H7, Salmonella enterica, and Listeria monocytogenes in acidified tomato-based foods. *Food Protection Trends, 37*(6), 409–418.

Endersen, L., Buttimer, C., Nevin, E., Coffey, A., Neve, H., Oliveira, H., et al. (2017). Investigating the biocontrol and anti-biofilm potential of a three phage cocktail against Cronobacter sakazakii in different brands of infant formula. *International Journal of Food Microbiology, 253*, 1–11. https://doi.org/10.1016/j.ijfoodmicro.2017.04.009.

Eser Öner, M., 2017. Atımlı Işık Uygulaması ile Bademde Salmonella Enteritidis İnaktivasyonu. Akademik Gıda, pp.242–248. https://doi.org/10.24323/akademik-gida.345257

Fernández, A., & Thompson, A. (2012). The inactivation of Salmonella by cold atmospheric plasma treatment. *Food Research International, 45*(2), 678–684. https://doi.org/10.1016/j.foodres.2011.04.009.

FINE, F., & GERVAIS, P. (2004). Efficiency of pulsed UV light for microbial decontamination of food powders. *Journal of Food Protection, 67*(4), 787–792. https://doi.org/10.4315/0362-028x-67.4.787.

Forghani, F., den Bakker, M., Liao, J.-Y., Payton, A. S., Futral, A. N., & Diez-Gonzalez, F. (2019). Salmonella and Enterohemorrhagic Escherichia coli serogroups O45, O121, O145 in wheat flour: Effects of long-term storage and thermal treatments. *Frontiers in Microbiology, 10*, 323. Available at: https://pubmed.ncbi.nlm.nih.gov/30853953.

Garud, S. R., Negi, P. S., & Rastogi, N. K. (2019). Improving the efficacy of ozone treatment in food preservation. In *Non-thermal Processing of Foods* (pp. 213–233). https://doi.org/10.1201/b22017-12.

Go, S.-M., Park, M.-R., Kim, H.-S., Choi, W. S., & Jeong, R.-D. (2019). Antifungal effect of non-thermal atmospheric plasma and its application for control of postharvest Fusarium oxysporum decay of paprika. *Food Control, 98*, 245–252. https://doi.org/10.1016/j.foodcont.2018.11.028.

González, S., Flick, G. J., Arritt, F. M., Holliman, D., & Meadows, B. (2006). Effect of high-pressure processing on strains of Enterobacter sakazakii. *Journal of Food Protection, 69*(4), 935–937. https://doi.org/10.4315/0362-028x-69.4.935.

Gurtler, J. B., Doyle, M. P., & Kornacki, J. L. (2014). The microbiological safety of spices and low-water activity foods: Correcting historic misassumptions. In *The Microbiological Safety of Low Water Activity Foods and Spices* (pp. 3–13). New York: Springer. https://doi.org/10.1007/978-1-4939-2062-4_1.

Haughton, P., Garvey, M., & Rowan, N. J. (2010). Emergence of Bacillus Cereus as a dominant organism in IRISH retailed powdered infant formulae (PIF) when reconstituted and stored under abuse conditions. *Journal of Food Safety, 30*(4), 814–831. https://doi.org/10.1111/j.1745-4565.2010.00244.x.

Hertwig, C., Reineke, K., Ehlbeck, J., Erdoğdu, B., Rauh, C., & Schlüter, O. (2015b). Impact of remote plasma treatment on natural microbial load and quality parameters of selected herbs and spices. *Journal of Food Engineering, 167*, 12–17. https://doi.org/10.1016/j.jfoodeng.2014.12.017.

Hertwig, C., Reineke, K., Ehlbeck, J., Knorr, D., & Schlüter, O. (2015a). Decontamination of whole black pepper using different cold atmospheric pressure plasma applications. *Food Control, 55*, 221–229. https://doi.org/10.1016/j.foodcont.2015.03.003.

Hu, L., Ma, L. M., Zheng, S., He, X., Wang, H., Brown, E. W., et al. (2017). Evaluation of 3M molecular detection system and ANSR pathogen detection system for rapid detection of Salmonella from egg products. *Poultry Science, 96*(5), 1410–1418. https://doi.org/10.3382/ps/pew399.

Humphries, R. M., & Linscott, A. J. (2015). Laboratory diagnosis of bacterial gastroenteritis. *Clinical microbiology reviews, 28*(1), 3–31. Available at: https://pubmed.ncbi.nlm.nih.gov/25567220.

Hwang, H.-J., Cheigh, C.-I., & Chung, M.-S. (2017). Construction of a pilot-scale continuous-flow intense pulsed light system and its efficacy in sterilizing sesame seeds. *Innovative Food Science & Emerging Technologies, 39*, 1–6. https://doi.org/10.1016/j.ifset.2016.10.017.

Jones, G., Pardos de la Gandara, M., Herrera-Leon, L., Herrera-Leon, S., Varela Martinez, C., Hureaux-Roy, R., et al. (2019). Outbreak of Salmonella enterica serotype Poona in infants linked to persistent Salmonella contamination in an infant formula manufacturing facility, France, August 2018 to February 2019. *Euro Surveillance, 24*(13), 1900161. Available at: https://pubmed.ncbi.nlm.nih.gov/30940315.

Jun, S., Irudayaraj, J., Demirci, A., & Geiser, D. (2003). Pulsed UV-light treatment of corn meal for inactivation of Aspergillus Niger spores. *International Journal of Food Science and Technology, 38*(8), 883–888. https://doi.org/10.1046/j.0950-5423.2003.00752.x.

Juneja, V. K., Friedman, M., Mohr, T. B., Silverman, M., & Mukhopadhyay, S. (2017). Control of Bacillus cereus spore germination and outgrowth in cooked rice during chilling by nonorganic and organic apple, orange, and potato peel powders. *Journal of Food Processing and Preservation, 42*(3), e13558. https://doi.org/10.1111/jfpp.13558.

Kao, Y.-T., Saxena, C., Wang, L., Sancar, A., & Zhong, D. (2005). Direct observation of thymine dimer repair in DNA by photolyase. *Proceedings of the National Academy of Sciences of the United States of America, 102*(45), 16128–16132. Available at: https://pubmed.ncbi.nlm.nih.gov/16169906.

Kazi, M., Parlapani, F. F., Boziaris, I. S., Vellios, E. K., & Lykas, C. (2017). Effect of ozone on the microbiological status of five dried aromatic plants. *Journal of the Science of Food and Agriculture, 98*(4), 1369–1373. https://doi.org/10.1002/jsfa.8602.

Keith, W. D., Harris, L. J., Hudson, L., & Griffiths, M. W. (1997). Pulsed electric fields as a processing alternative for microbial reduction in spice. *Food Research International, 30*(3–4), 185–191. https://doi.org/10.1016/s0963-9969(97)00028-8.

Kim, J. E., Lee, D.-U., & Min, S. C. (2014). Microbial decontamination of red pepper powder by cold plasma. *Food Microbiology, 38*, 128–136. https://doi.org/10.1016/j.fm.2013.08.019.

Kim, J. E., Oh, Y. J., Won, M. Y., Lee, K.-S., & Min, S. C. (2017). Microbial decontamination of onion powder using microwave-powered cold plasma treatments. *Food Microbiology, 62*, 112–123. https://doi.org/10.1016/j.fm.2016.10.006.

Kim, J. E., Oh, Y. J., Song, A. Y., & Min, S. C. (2018). Preservation of red pepper flakes using microwave-combined cold plasma treatment. *Journal of the Science of Food and Agriculture, 99*(4), 1577–1585. https://doi.org/10.1002/jsfa.9336.

Kim, J. H., & Min, S. C. (2018). Moisture vaporization-combined helium dielectric barrier discharge-cold plasma treatment for microbial decontamination of onion flakes. *Food Control, 84*, 321–329. https://doi.org/10.1016/j.foodcont.2017.08.018.

Kim, M.-J., Han, J., Park, J.-S., Lee, J.-S., Lee, S.-H., Cho, J.-I., & Kim, K.-S. (2015). Various enterotoxin and other virulence factor genes widespread among Bacillus cereus and Bacillus thuringiensis strains. *Journal of Microbiology and Biotechnology, 25*(6), 872–879. https://doi.org/10.4014/jmb.1502.02003.

Koch, F., Wiacek, C., & Braun, P. G. (2019). Pulsed light treatment for the reduction of Salmonella typhimurium and Yersinia enterocolitica on pork skin and pork loin. *International Journal of Food Microbiology, 292*, 64–71. https://doi.org/10.1016/j.ijfoodmicro.2018.11.014.

Lee, H., Ha, M. J., Shahbaz, H. M., Kim, J. U., Jang, H., & Park, J. (2018a). High hydrostatic pressure treatment for manufacturing of red bean powder: A comparison with the thermal treatment. *Journal of Food Engineering, 238*, 141–147. https://doi.org/10.1016/j.jfoodeng.2018.06.016.

Lee, S. J., Bang, I. H., Choi, H.-J., & Min, S. C. (2018b). Pasteurization of mixed mandarin and Hallabong tangor juice using pulsed electric field processing combined with heat. *Food Science and Biotechnology, 27*(3), 669–675. Available at: https://pubmed.ncbi.nlm.nih.gov/30263793.

Lehmacher, A., Bockemühl, J., & Aleksic, S. (1995). Nationwide outbreak of human salmonellosis in Germany due to contaminated paprika and paprika-powdered potato chips. *Epidemiology and Infection, 115*(3), 501–511. Available at: https://pubmed.ncbi.nlm.nih.gov/8557082.

Li, R., Fei, P., Man, C. X., Lou, B. B., Niu, J. T., Feng, J., et al. (2016). Tea polyphenols inactivate Cronobacter sakazakii isolated from powdered infant formula. *Journal of Dairy Science, 99*(2), 1019–1028. https://doi.org/10.3168/jds.2015-10039.

Lin, Y.-D., & Chou, C.-C. (2004). Effect of heat shock on thermal tolerance and susceptibility of Listeria monocytogenes to other environmental stresses. *Food Microbiology, 21*(5), 605–610. https://doi.org/10.1016/j.fm.2003.10.007.

Marco, A., Ferrer, C., Velasco, L. M., Rodrigo, D., Muguerza, B., & Martínez, A. (2011). Effect of olive powder and high hydrostatic pressure on the inactivation of Bacillus cereus spores in a reference medium. *Foodborne Pathogens and Disease, 8*(6), 681–685. https://doi.org/10.1089/fpd.2010.0712.

Morales-de la Peña, M., Salvia-Trujillo, L., Rojas-Graü, M. A., & Martín-Belloso, O. (2010). Impact of high intensity pulsed electric field on antioxidant properties and quality parameters of a fruit juice–soymilk beverage in chilled storage. *LWT - Food Science and Technology, 43*(6), 872–881. https://doi.org/10.1016/j.lwt.2010.01.015.

Moreau M Nicorescua, I., Turpina, A., Agoulonb, A., Chevaliera, S., & Orangea, N. (2009) Decontamination of spices by using a pulsed light treatment In: Food process engineering in a changing world. (1997)

Mošovská, S., Medvecká, V., Halászová, N., Ďurina, P., Valík, Ľ., Mikulajová, A., & Zahoranová, A. (2018). Cold atmospheric pressure ambient air plasma inhibition of pathogenic bacteria on the surface of black pepper. *Food Research International, 106*, 862–869. https://doi.org/10.1016/j.foodres.2018.01.066.

Nicorescu, I., Nguyen, B., Moreau-Ferret, M., Agoulon, A., Chevalier, S., & Orange, N. (2013). Pulsed light inactivation of Bacillus subtilis vegetative cells in suspensions and spices. *Food Control, 31*(1), 151–157. https://doi.org/10.1016/j.foodcont.2012.09.047.

Oey, I., Lille, M., Van Loey, A., & Hendrickx, M. (2008). Effect of high-pressure processing on colour, texture and flavour of fruit- and vegetable-based food products: A review. *Trends in Food Science & Technology, 19*(6), 320–328. https://doi.org/10.1016/j.tifs.2008.04.001.

Oonaka, K., Furuhata, K., Hara, M., & Fukuyama, M. (2010). Powder infant formula milk contaminated with Enterobacter sakazakii. *Japanese Journal of Infectious Diseases, 63*(2), 103–107.

Park, I., Kim, J. U., Shahbaz, H. M., Jung, D., Jo, M., Lee, K. S., et al. (2018). High hydrostatic pressure treatment for manufacturing of garlic powder with improved microbial safety and antioxidant activity. *International Journal of Food Science & Technology, 54*(2), 325–334. https://doi.org/10.1111/ijfs.13937.

Patil, S., Moiseev, T., Misra, N. N., Cullen, P. J., Mosnier, J. P., Keener, K. M., & Bourke, P. (2014). Influence of high voltage atmospheric cold plasma process parameters and role of relative humidity on inactivation of Bacillus atrophaeus spores inside a sealed package. *Journal of Hospital Infection, 88*(3), 162–169. https://doi.org/10.1016/j.jhin.2014.08.009.

Pina-Pérez, M. C., Martínez-López, A., & Rodrigo, D. (2013b). Cocoa powder as a natural ingredient revealing an enhancing effect to inactivate Cronobacter sakazakii cells treated by pulsed electric fields in infant milk formula. *Food Control, 32*(1), 87–92. https://doi.org/10.1016/j.foodcont.2012.11.014.

Pina-Pérez, M. C., Rodrigo, D., & Martínez, A. (2016). Nonthermal inactivation of Cronobacter sakazakii in infant formula milk: A review. *Critical Reviews in Food Science and Nutrition, 56*(10), 1620–1629.

Pina-Pérez, M. C., Rodrigo, D., & Martínez-López, A. (2011). Bacteriostatic effect of cocoa powder rich in polyphenols to control Cronobacter sakazakii proliferation on infant milk formula. *Science and Technology Against Microbial Pathogens*. https://doi.org/10.1142/9789814354868_0016.

Pina-Pérez, M. C., Silva-Angulo, A. B., Rodrigo, D., & Martínez López, A. (2012). A preliminary exposure assessment model for Bacillus cereus cells in a milk based beverage: Evaluating high pressure processing and antimicrobial interventions. *Food Control, 26*(2), 610–613. https://doi.org/10.1016/j.foodcont.2012.01.063.

Pina-Pérez, M. C., Benlloch-Tinoco, M., Rodrigo, D., & Martinez, A. (2013a). Cronobacter sakazakii inactivation by microwave processing. *Food and Bioprocess Technology, 7*(3), 821–828. https://doi.org/10.1007/s11947-013-1063-2.

Pina-Pérez, M. C., Rivas, A., Martínez, A., & Rodrigo, D. (2018). Effect of thermal treatment, microwave, and pulsed electric field processing on the antimicrobial potential of açaí (Euterpe oleracea), stevia (Stevia rebaudiana Bertoni), and ginseng (Panax quinquefolius L.) extracts. *Food Control, 90*, 98–104. https://doi.org/10.1016/j.foodcont.2018.02.022.

Podolak, R., Enache, E., Stone, W., Black, D. G., & Elliott, P. H. (2010). Sources and risk factors for contamination, survival, persistence, and heat resistance of Salmonella in low-moisture foods. *Journal of Food Protection, 73*(10), 1919–1936. https://doi.org/10.4315/0362-028x-73.10.1919.

Proctor, A. D., Ahmedna, M., Kumar, J. V, & Goktepe, I. (2004). Degradation of aflatoxins in peanut kernels/flour by gaseous ozonation and mild heat treatment. *Food Additives and Contaminants, 21*(8), 786–793. https://doi.org/10.1080/02652030410001713898.

Reilly, A. Tlustos, C., O'Connor, J., & O'Connor, L., (2009). Food safety: a public health issue of growing importance. In: *Introduction to human nutrition*, p 324

Roohinejad, S., Koubaa, M., Sant'Ana, A. S., & Greiner, R. (2018). Mechanisms of microbial inactivation by emerging technologies. *Innovative Technologies for food Preservation, 2018*, 111–132. https://doi.org/10.1016/b978-0-12-811031-7.00004-2.

Shi, C., Sun, Y., Liu, Z., Guo, D., Sun, H., Sun, Z., et al. (2017). Inhibition of Cronobacter sakazakii virulence factors by citral. *Scientific Reports, 7*, 43243. Available at: https://pubmed.ncbi.nlm.nih.gov/28233814.

Stojiljkovic, J., Trajcev, M., Petrovska, M., Petanovska, I., Trajkovski, B., & G. (2016). The growth of Salmonella Enteritidis in egg-based pasta with the addition of sweet Basil and Thymus. *Cell Biology, 4*(6), 35–39.

Tiwari, B. K., Brennan, C. S., Curran, T., Gallagher, E., Cullen, P. J., & O' Donnell, C. P. (2010). Application of ozone in grain processing. *Journal of Cereal Science, 51*(3), 248–255. https://doi.org/10.1016/j.jcs.2010.01.007.

Tsujimoto, H., Huang, C. C., Kinoshita, N., Inoue, Y., Eitoku, H., & Sekiguchi, I. (2004). Ultrahigh pressure sterilization of powdery food stuff—a new application of a roller compactor. *Powder Technology, 146*(3), 214–222. https://doi.org/10.1016/j.powtec.2004.08.009.

Vega-Mercado, H., Martín-Belloso, O., Chang, F.-J., Barbosa-Ccanovas, G. V., & Swanson, B. G. (1996). Inactivation of escherichia coli and bacillus subtilis suspended in pea soup using pulsed electric fields. *Journal of Food Processing and Preservation, 20*(6), 501–510. https://doi.org/10.1111/j.1745-4549.1996.tb00762.x.

Wang, B., Liu, S., Sui, Z., Wang, J., Wang, Y., & Gu, S. (2020). Rapid flow cytometric detection of single viable Salmonella cells in Milk powder. *Foodborne Pathogens and Disease*. https://doi.org/10.1089/fpd.2019.2748.

Windyga, B., Fonberg-Broczek, M., Sciezyńska, H., Skapska, S., Górecka, K., Grochowska, A., et al. (2008). High pressure processing of spices in atmosphere of helium for decrease of microbiological contamination. *Roczniki Państwowego Zakładu Higieny, 59*(4), 437–443.

Xin, W., Huang, Y., Ji, B., Li, P., Wu, Y., Liu, J., et al. (2019). Identification and characterization of Clostridium botulinum strains associated with an infant botulism case in China. *Anaerobe, 55*, 1–7. https://doi.org/10.1016/j.anaerobe.2018.06.015.

Young, J. C., Zhu, H., & Zhou, T. (2006). Degradation of trichothecene mycotoxins by aqueous ozone. *Food and Chemical Toxicology, 44*(3), 417–424. https://doi.org/10.1016/j.fct.2005.08.015.

Yu, Y., Jin, T. Z., Fan, X., & Xu, Y. (2016). Osmotic dehydration of blueberries pretreated with pulsed electric fields: Effects on dehydration kinetics, and microbiological and nutritional qualities. *Drying Technology, 35*(13), 1543–1551. https://doi.org/10.1080/07373937.2016.1260583.

Yuste, J., Capellas, M., Pla, R., Fung, D. Y. C., & Mor-Mur, M. (2001). High Pressure processing for food safety and preservation: A review. *Journal of Rapid Methods & Automation in Microbiology, 9*(1), 1–10. https://doi.org/10.1111/j.1745-4581.2001.tb00223.x.

Zahoranová, A., Hoppanová, L., Šimončicová, J., Tučeková, Z., Medvecká, V., Hudecová, D., et al. (2018). Effect of cold atmospheric pressure plasma on maize seeds: Enhancement of seedlings growth and surface microorganisms inactivation. *Plasma Chemistry and Plasma Processing, 38*(5), 969–988. https://doi.org/10.1007/s11090-018-9913-3.

Zhang, J., Davis, T. A., Matthews, M. A., Drews, M. J., LaBerge, M., & An, Y. H. (2006). Sterilization using high-pressure carbon dioxide. *The Journal of Supercritical Fluids, 38*(3), 354–372. https://doi.org/10.1016/j.supflu.2005.05.005.

Index

E. Ermiş (ed.), *Food Powders Properties and Characterization*, Food Engineering Series, https://doi.org/10.1007/978-3-030-48908-3